AXON
ALGEBRA 1

Challenge and Enrichment Masters

SAXON™

An Imprint of HMH
Supplemental Publishers

www.SaxonPublishers.com
1-800-531-5015

ISBN 13: 978-1-6027-7492-6

ISBN 10:   1-6027-7492-7

1 2 3 4 5 6 7 8   170   15 14 13 12 11 10 09 08

Contents

Challenge and Enrichment

Challenges 1 and 2..1
Challenges 3 and 4..2
Challenges 5 and 6..3
Challenges 7 and 8..4
Challenges 9 and 10..5
Challenges 11 and 12..6
Challenges 13 and 14..7
Challenges 15 and 16..8
Challenges 17 and 18..9
Challenges 19 and 20..10
Challenges 21 and 22..11
Challenges 23 and 24..12
Challenges 25 and 26..13
Challenges 27 and 28..14
Challenges 29 and 30..15
Challenges 31 and 32..16
Challenges 33 and 34..17
Challenges 35 and 36..18
Challenges 37 and 38..19
Challenges 39 and 40..20
Challenges 41 and 42..21
Challenges 43 and 44..22
Challenges 45 and 46..23
Challenges 47 and 48..24
Challenges 49 and 50..25
Challenges 51 and 52..26
Challenges 53 and 54..27
Challenges 55 and 56..28
Challenges 57 and 58..29
Challenges 59 and 60..30
Challenges 61 and 62..31
Challenges 63 and 64..32
Challenges 65 and 66..33
Challenges 67 and 68..34
Challenges 69 and 70..35
Challenges 71 and 72..36
Challenges 73 and 74..37
Challenges 75 and 76..38
Challenges 77 and 78..39
Challenges 79 and 80..40
Challenges 81 and 82..41
Challenges 83 and 84..42
Challenges 85 and 86..43
Challenges 87 and 88..44

       Saxon Algebra 1

Challenges 89 and 90...45

Challenges 91 and 92...46

Challenges 93 and 94...47

Challenges 95 and 96...48

Challenges 97 and 98...49

Challenges 99 and 100..50

Challenges 101 and 102..51

Challenges 103 and 104..52

Challenges 105 and 106..53

Challenges 107 and 108..54

Challenges 109 and 110..55

Challenges 111 and 112..56

Challenges 113 and 114..57

Challenges 115 and 116..58

Challenges 117 and 118..59

Challenges 119 and 120..60

Enrichment 3 61

Enrichment 4 62

Enrichment 7 63

Enrichment 9 64

Enrichment 11...65

Enrichment 13...66

Enrichment 17...67

Enrichment 20...68

Enrichment 23...69

Enrichment 25...70

Enrichment 26...71

Enrichment 28...72

Enrichment 30...73

Enrichment 31...74

Enrichment 42...75

Enrichment 44...76

Enrichment 45...77

Enrichment 58...78

Enrichment 60...79

Enrichment 64...80

Enrichment 67...81

Enrichment 72...82

Enrichment 75...83

Enrichment 82...84

Enrichment 83...85

Enrichment 85...86

Enrichment 93...87

Enrichment 104..88

Enrichment 105..89

Enrichment 109..90

Enrichment 115..91

Answer Key 92

 Saxon Algebra 1

Challenge
Classifying Real Numbers

1

Challenge

Tyrone runs 7 laps on a quarter-mile track. Determine whether or not the distance Tyrone runs during practice would remain in the same subset of real numbers if he runs 8, 9, or 10 laps. If not, identify any additional subsets.

- -

Challenge
Understanding Variables and Expressions

2

Challenge

The variable h represents the number of hours. To find the charges for a bike rental for 3 hours, let $h = 3$. In other words, the charge for 3 hours is $5 + $2.25 · 3. Identify the variables and constants in this expression.

What will the bicycle store charge for 3 hours?

Name _____ Date _____ Class _____

Challenge

Simplify $\left(\frac{2}{3}\right)^2\left(\frac{2}{3}\right)^2\left(\frac{9}{4}\right)\left(\frac{9}{4}\right)$ using the Product Rule of Exponents and then check using the order of operations. Show your work.

- -

Name _____ Date _____ Class _____

Challenge

Add parentheses in the equations below to make the number sentences true.

  a. $9 + 12 \div 3 - 1 = 15$

  b. $3 \cdot 4^2 - 2^3 + 3^2 = 23$

Saxon Algebra 1

Challenge 5
Finding Absolute Value and Adding Real Numbers

Challenge

Simplify.

$-3 + |-5|-7- 4 - \left|(-2) + |1 + (-3)|\right|$

- -

Challenge 6
Subtracting Real Numbers

Challenge

Simplify the expression.

$25 - (-28) - [57 - (-86 - 19)]$

Challenge **7**
Simplifying and Comparing Expressions with Symbols of Inclusion

Challenge

Grouping symbols can be added in a variety of places to change the value of a numeric expression. Insert grouping symbols into the following expression to create several different values.

$3 + 4 \cdot 6^2 - 5 \cdot 7 + 11$

Challenge **8**
Using Unit Analysis to Convert Measures

Challenge

Many unit analysis problems include the conversion of more than one unit. For example, a unit of speed may be converted from miles per hour to feet per second. Convert 44 miles per hour to feet per second.

4   **Saxon** Algebra 1

Challenge 9
Evaluating and Comparing Algebraic Expressions

Challenge

Sometimes an expression will include more than two variables. Evaluate the expression $2(z - x)^2 + 2v^2 - 2wy$ for $v = 4$, $w = 9$, $x = 13$, $y = 3$, and $z = 20$.

- -

Challenge 10
Adding and Subtracting Real Numbers

Challenge

Order from greatest to the least.

$(-0.30 - 0.02)$, 0.34, $\dfrac{1}{3}$, $\left(-\dfrac{1}{6} - \dfrac{1}{6}\right)$

Saxon Algebra 1

Name _____ Date _____ Class _____

Challenge

Explain how to simplify the expression $-4(-2)^3$.

- -

Name _____ Date _____ Class _____

Challenge

Explain why the Identity Property applies to subtraction but not to the Commutative Property.

     Saxon Algebra 1

Challenge **13**
Calculating and Comparing Square Roots

Challenge

Compare the expressions using estimation.

$\sqrt{800} \;\bigcirc\; \sqrt{170} + \sqrt{330}$

- -

Challenge **14**
Determining the Theoretical Probability of an Event

Challenge

The odds in favor of an event is the ratio of the probability of the event to the probability of its complement. If the event is A, and its complement is $\overline{A}$, then the odds for A are $P(A) : P(\overline{A})$ or $\dfrac{P(A)}{P(\overline{A})}$.

Find the odds of an event with 2 equally likely outcomes.

Saxon Algebra 1

Challenge 15
Using the Distributive Property to Simplify Expressions

Challenge

Simplify the expression:

$(x - 2y)(m + y)$

Challenge 16
Simplifying and Evaluating Variable Expressions

Challenge

Square roots and cubic roots can also be expressed as fractional exponents. For example,

$\sqrt{x} = x^{\frac{1}{2}}$ and $\sqrt[3]{x} = x^{\frac{1}{3}}$

Evaluate the following expression for the given values of the variables:

$[(x^{\frac{1}{2}} + y)z]^{\frac{1}{3}}$ for $x = 9$, $y = 2$, $z = 25$

Saxon Algebra 1

Challenge **17**
Translating Between Words and Algebraic Expressions

Challenge

For example, $18 - (4 + x)$ translates to "eighteen minus the quantity four plus x."

Translate the following phrases:

"the product of five and the quantity two less than a number"

"the difference of seven and the quantity three times a number plus two"

- -

Challenge **18**
Combining Like Terms

Challenge

Use the reverse of the Distributive Property to divide out constants and variables from $3x^2y - 4xy$. Then distribute the factor to check your work.

Saxon Algebra 1

Challenge

Solve.

$3x + 3 - 2x = -8$

$x + 4 - 9 = 6$

$x - 4 = 9 + (-5)$

$2x + 3x - 2 - 4x = 7$

$-5 + x - 4 = 12$

— —

Challenge

Create a square on a coordinate plane by having all of the coordinates be in a different quadrant. Name each coordinate.

Create other geometric figures.

Challenge **21**
Solving One-Step Equations by Multiplying or Dividing

Challenge

Sometimes you have to use more than one equation to find the answer to a problem. The combined area of 4 equal-sized squares is 324 ft^2. What is the perimeter of each square? Write and solve equations to find the answer.

- -

Challenge **22**
Analyzing and Comparing Statistical Graphs

Challenge

The table shows Andre's bank account transactions.

Month	January	February	March	April	May	June
Deposits	$475	$200	$350	$425	$500	$150
Withdrawals	$100	$275	$350	$400	$200	$225

Make a graph that displays his account balance at the end of each month.

(Assume there was no money in Andre's account before January.)

     Saxon Algebra 1

Challenge

Solve.

$5x - 6 - 2x = 3$

- -

Challenge

In some problems, you need to find the decimal part of a decimal number.
Solve and check that your answer is reasonable.

0.35 of 12.6 is what number?

Challenge

Division by zero is undefined. Sometimes certain values cannot be used for the domain of a function because they will cause the expression to be undefined. Tell which values cannot be used for the domain of each function and why.

$f(x) = \dfrac{2}{x}$, $g(x) = \dfrac{x + 1}{x - 6}$, $h(x) = \dfrac{1}{(x + 1)^2}$

— —

Challenge

An equation can have more than one set of grouping symbols. Start by using the Distributive Property on the innermost set of grouping symbols and work outward.

Solve: $3[x - (2x + 1)] + 4(x + 7) = -100$

Saxon Algebra 1

Challenge

The following data shows the average gas price of regular gasoline per gallon over a time span of 7 years.

1999: $1.17; 2000: $1.51; 2001: $1.46; 2002: $1.36; 2003: $1.59; 2004: $1.88; 2005: $2.30

Create two different graphs of the data. One graph would imply that gasoline prices skyrocketed during the 7 years, and the other would make it seem that there was little change in gasoline prices over the same period. Then decide who might make each type of graph and why.

- -

Challenge

Solve the equation by first multiplying by a reciprocal. Show each step.

$8(2x - 1) + 15 + 7x = 4(3x + 5) + 5x - 1$

Saxon Algebra 1

Challenge **29**
Solving Literal Equations

Challenge

The formula for the area of a circle is $A = \pi r^2$ If the circle's area measures
81 m^2, what is the radius? Leave the symbol π in the answer.

- -

Challenge **30**
Graphing Functions

Challenge

Make a table of values and graph of the equation
$y = x^2 - 2x$. Is the graph a function? Is it linear?

Challenge

The ratio of four times a number to the sum of the same number and three is 5:2. What is the number?

- -

Challenge

Simplify the expression.

$$\frac{\dfrac{a^{-2}b^{6}c^{-3}}{a^{4}b^{8}c^{-5}}}{b^{-5}}$$

Challenge **33**
Finding the Probability of Independent and Dependent Events

Challenge

A bag of shapes has 3 circles, 2 squares, and 4 triangles. Three shapes are drawn at random from the bag. The first shape is drawn and is not returned to the bag; the second shape is drawn and is returned to the bag. What is the probability of drawing the shapes in the following order: square, circle, triangle? Show your work.

- -

Challenge **34**
Recognizing and Extending Arithmetic Sequences

Challenge

The formula represents the sum of the first n terms of an arithmetic sequence with first term a_1 and nth term a_n:

$$S_n = \frac{n}{2}(a_1 + a_n)$$

Find the sum of the positive integers less than 100 and divisible by 6. Explain your solution.

     Saxon Algebra 1

Challenge

Describe the graph of a line drawn through the following points:
$(5, -2)$, $(5, 3)$. Draw a line through the points and determine the x- and
y-intercepts. Write an equation in standard form representing the graph of
the line.

- -

Challenge

If a triangle has sides lengths x, y, and z, and a similar triangle has
lengths $2.5x$, $2.5y$, $2.5z$, what is the ratio of the area of the triangle to the
similar triangle?

Name _____ Date _____ Class _____

Challenge

Write each number in scientific notation and then simplify the expression.

$$\frac{\dfrac{35,000,000}{20,000} \times \dfrac{81,000}{15,000}}{16,000,000 \times 74,000}$$

- -

Name _____ Date _____ Class _____

Challenge

Write a 4-term expression that can be factored using the GFC $6g^4h^7k^2$.
Then factor the expression you wrote.

Saxon Algebra 1

Challenge

Some problems require distributing more than one factor. Simplify the expression.

$$\left(\frac{ab^{-2}}{c^2} + 4b^{-1}d^{-2}\right)\left(\frac{a^{-3}db^4}{c^{-7}} - \frac{5b^{-2}}{c^{-3}}\right)$$

– –

Challenge

a. Fill in the boxes with values that will make the equation true.

$$(x^9y^6z^3) = (x^\square y^\square z^\square)^{-3}$$

b. Simplify $(x^{-4}y^2z^6)^2(x^2y^{-1}z^{-3})^4$.

Challenge **41**
Finding Rates of Change and Slope

Challenge

Graph the line that passes through the point $(-2, 3)$ and has a slope of -3. (Hint: Write -3 as a fraction.)

- -

Challenge **42**
Solving Percent Problems

Challenge

An after-school program received a grant for $21,327. Program administrators want to use $\frac{1}{4}$ of the funds to pay tutors, 0.5 of the funds for a new facility, 12% of the funds for a parent-student program, and 13% of the funds for materials. Estimate the amount of funds available for each category.

Challenge

Use guess and check to find undefined values of the variables. Then simplify the rational expression.

$$\frac{x^2 - 1}{2x^2 - 2}$$

- -

Challenge

Car A traveled a father distance than Car B over the same amount of time. Which car traveled at a faster speed? Explain.

Challenge 45
Translating Between Words and Inequalities

Challenge

Translate these sentences into algebraic inequalities:

The absolute value of the difference of twice a number and 5 is not less than the square of the sum of 7 and the number.

Twice the square of the difference of the cube of a number and 1 is not more than the sum of half the number and 1.

- -

Challenge 46
Simplifying Expressions with Square Roots and Higher-Order Roots

Challenge

A gift box in the shape of a cube has a volume of 3375 cubic inches.

$V = 3375 \text{ in}^3$

Suppose a second gift box has a volume that is 8 times greater, so that $V = 8 \cdot 3375 \text{ in}^3$; the length of each side is $s = \sqrt[3]{8 \cdot 3375}$ inches. Without using a calculator, find the length of each side.

Challenge 47
Solving Problems Involving the Percent of Change

Challenge

Maria received a gift of shares in a company stock with a total value of $65. Fifteen years later, the stock is worth $868.39. What is the percent increase of the stock price?

Another stock that Maria owns fell in value from $74.08 to $73.63 in two years. What is the percent decrease?

— —

Challenge 48
Analyzing Measures of Central Tendency

Challenge

The following data represents the mean price per pound of ten different fresh fruits and vegetables in 2005.

$0.97, $0.48, $0.89, $1.10, $2.76, $1.51, $0.99, $0.50, $0.85, $1.85

Estimate the mean of the data by simply looking at the values.

How can you check your estimate with only addition and subtraction?

Challenge 49

Writing Equations in Slope-Intercept Form

Challenge

The ordered pairs (0, 0) and (1, 4) fall on the graph of an equation.
Determines the equation for the line and another point that lies on the line
within quadrant III.

- -

Challenge 50

Graphing Inequalities

Challenge

Marie is planning a bowling party. The bowling alley allows no more than
12 players in 2 lanes. Explain if an inequality would be appropriate to
represent this situation.

     Saxon Algebra 1

Name _____ Date _____ Class _____

Challenge **51**
Simplifying Rational Expressions with Like Denominators

Challenge

Write an expression that has more than one excluded value.

- -

Name _____ Date _____ Class _____

Challenge **52**
Determining the Equation of a Line Given Two Points

Challenge

Write the equation of a line that has no slope and an *x*-intercept of 22 in point-slope form.

Saxon Algebra 1

Challenge

A rectangular garden will be bordered by a stone walkway of width x.
Use the diagram below to find the area of the walkway for the purpose of purchasing the stones.

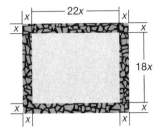

- -

Challenge

John Wilder Tukey (1915–2000) was an American statistician born in Massachusetts. Research this important modern mathematician by finding the statistical graph he invented, as well as the two computer terms he is credited with coining.

 Saxon Algebra 1

Name _____ Date _____ Class _____

Challenge

Sometimes one or both equations in a system will be nonlinear. The solution(s) to these systems will still be at the points where the graphs intersect. Solve this system by graphing:

$y = x$
$x^2 + y^2 = 8$

- -

Name _____ Date _____ Class _____

Challenge

Draw two lines with different slopes that intersect at the origin and write the equation for those lines. Then write a word problem that involves the two direct variations.

Explain the significance of the graphs either passing through or starting at (0, 0).

Challenge 57
Finding the Least Common Multiple

Challenge

Find the LCM of
$22x^{-2}x^4yz^3$ and $26xy^{-3}y^5z$ and $34x^3y^3z^3$.

- -

Challenge 58
Multiplying Polynomials

Challenge

Find these products mentally. You may scribble notes on scratch paper, but actually writing the problem is not permitted.

$(m + 3)(m - 4)$

$(3p - 8)(5p + 4)$

$(2a + 1)(3a^2 - 2a + 7)$

Saxon Algebra 1

Challenge

Use the substitution method to solve the following system. Then graph the system and explain the relationship between the algebraic result and the graph.

$$y = x + 2$$
$$2y - 2x = 4$$

- -

Challenge

Factoring is the inverse of multiplying. Use your knowledge of special products to write each of the following expressions as a product of two binomials:

a. $x^2 - 64$

b. $4x^2 - 25$

c. $9a^2 - 16b^2$

d. $x^2 + 14x + 49$

e. $4a^2 + 12a + 9$

f. $9x^2 + 42xy + 49y^2$

 Saxon Algebra 1

Challenge

Simplify.

a. $\sqrt[3]{8g^3h^4}$

b. $\sqrt[5]{32y^3z^8}$

– –

Challenge

Create your own survey and collect data. Then create a stem-and-leaf plot and histogram, find measures of center and spread, and discuss the distribution of the data based on the data displays that you created.

   Saxon Algebra 1

Challenge

Use elimination to solve for x, y, and z in the following system of equations.

$$-x + 2y + 3z = 5$$
$$3y + 2z + 1 = x$$
$$z - x = 1$$

- -

Challenge

A number z varies directly with y and indirectly with x. When $y = 14$ and $x = 7$, $z = 8$. Write the equation that models this relationship. Then find z when $y = 32$ and $x = 8$.

Challenge

A rhombus is a quadrilateral with two pairs of parallel sides and perpendicular diagonals. Given that quadrilateral *QRST* has vertices at $Q(1, 4)$, $R(2, 1)$, $S(-1, 0)$, and $T(-2, 3)$, show that *QRST* is a rhombus.

— —

Challenge

Sometimes a variable appears on both sides of an inequality. Determine which inequalities below are true by substituting values for *x*.

$x > x$

$x \le x$

$x - 1 < x$

$x - 3 \ge x - 2$

Challenge **67**
Solving and Classifying Special Systems of Linear Equations

Challenge

Write a system of equations in standard form with the given classification.

a. inconsistent

b. consistent and dependent

c. consistent and independent

- -

Challenge **68**
Mutually Exclusive and Inclusive Events

Challenge

Suppose Mariam needs to survey 200 homeowners to find out if they prefer paint or vinyl siding. Assume that of a local population of 320,000, 100,000 live in rented apartments, and 50,000 homeowners live in brick or stucco homes and have no preference for paint or vinyl siding. About how people can Mariam expect to call who prefer paint or vinyl siding?

34 **Saxon** Algebra 1

Challenge

Find the missing radicands in these true statements.

a. $\sqrt{\square} - \sqrt{3} = \sqrt{3}$

b. $-2\sqrt{\square} = -6\sqrt{x}$

c. $3x\sqrt{x} + \sqrt{\square} = 5x\sqrt{x}$

- -

Challenge

Formulate and solve an inequality for the problem below.

At least 30% of a solution must be salt to form a usable solution. If there is only 7.8 grams of salt, then how many grams of solution can be formed?

Saxon Algebra 1

Challenge 71
Making and Analyzing Scatter Plots

Challenge

Even though a strong correlation may exist between sets of data values, one set does not necessarily cause the other. A positive correlation may exist between time spent studying and test grades, but time spent studying may be only one cause of the outcome. Explain what might be the reason for the positive correlations given.

   a. time spent studying and test grades

   b. the sale of swimsuits and the outside temperature

   c. the number of fans at a soccer game and the outside temperature

— —

Challenge 72
Factoring Trinomials: $x^2 + bx + c$

Challenge

Factor $x^2 - 2x - 168$.

Saxon Algebra 1

Challenge

Conjunctions can involve more than two inequalities. Although more complicated, the method to solving these compound inequalities is still the same. Solve this system:

$2x - 3 > 3$ AND $-x + 3 > -4$ AND $3x - 4 \le 11$

_ _

Challenge

Translate and solve the following statement: The distance from x to 3 times 2 plus 4 is 14.

Challenge

Factor.

1. $9p^2s - 30ps^2 + 25s^3$

2. $6k^3 + 5k^2 + k$

3. $30m^2 - 87m + 30$

4. $9a^3b - 24a^2b^2 + 16ab^3$

5. $42x^3 + 45x^2 - 27x$

Challenge

Use the FOIL method to multiply $\left(3x + \sqrt{5}\right)\left(3x - \sqrt{5}\right)$.

Challenge **77**
Solving Two-Step and Multi-Step Inequalities

Challenge

Solve.

1. $3[4(b - 2) - (1 - b)] > 5(b - 4)$

2. $\frac{2}{3}(9m - 15) + 4 < 6 + \frac{3}{4}(4 - 12m)$

- -

Challenge **78**
Graphing Rational Functions

Challenge

$$y = \frac{(x + 1)(x - 3)}{(x - 5)(x + 2)}$$

Describe the graph of this function without graphing it. Check your predictions with a graphing calculator.

Saxon Algebra 1

Challenge

The greatest common factor of the terms of a binomial may be another binomial. Example: $x(x + 5) + 4(x + 5) = (x + 5)(x + 4)$.

Factor these binomials.

  a. $(xz - yz) + (2x - 2y)$

  b. $(q + pq) - (3 + 3p)$

- -

Challenge

The probability of an outcome is p. Let n represent the number of trials in this experiment. Write an algebraic expression to make a prediction about the outcome.

Challenge **81**
Solving Inequalities with Variables on Both Sides

Challenge

Determine if an ordered pair is a solution to a system of inequalities.

For $x > 0$ and $y \le 0$, is (4, 0) a solution?

For $x + 2 > 3$ and $3y < 9$, is (1, 1) a solution?

For $5x \le 24 - x$ and $3y < 8 + y$, is (−10, 4) a solution?

- -

Challenge **82**
Solving Multi-Step Compound Inequalities

Challenge

An average level of HDL for a person is no more than 60, and an unhealthy average level is lower than 40. Write a compound inequality that represents these averages.

Challenge

Factor $8x^4y^3 - 162x^2y$.

- -

Name _____ Date _____ Class _____

Challenge

Graph $y = x^2 + x + 1$.

Challenge **85**
Solving Problems Using the Pythagorean Theorem

Challenge

Develop a formula for finding other Pythagorean triples, using the
Pythagorean triple 3, 4, and 5.

- -

Challenge **86**
Calculating the Midpoint and Length of a Segment

Challenge

Find the perimeter of triangle *ABC* with vertices *A* (1, 2),
B (4, 5), and *C* (−7, −8). Round to the nearest hundredth.

43   **Saxon** Algebra 1

Challenge

Factor the polynomial completely.

$24x^3y - 28y^3 - 35x^2y^2 + 30x^5$

- -

Challenge

Simplify.

$$\frac{x^2 + 5x + 6}{x^2 + 13x + 42} \cdot \frac{x^2 + 8x + 7}{x^2 + 3x + 2}$$

 Saxon Algebra 1

Challenge

A farmer has 50 yards of fence to construct a rectangular kennel. The function $f(x) = 50x - 2x^2$ describes the areas that can be form with x width. Find the maximum area that can be constructed.

‒ ‒

Challenge

Simplify.

$$\frac{3x^6 y}{x^4 + 2x^2 y + y^2} + \frac{y}{x^5 y + x^3 y^2}$$

Challenge 91
Solving Absolute-Value Inequalities

Challenge

There are also compound absolute-value inequalities. Consider the inequalities:

$|x| < 6$ AND $|x - 2| > 6$

Solve the compound absolute-value inequalities.

- -

Challenge 92
Simplifying Complex Fractions

Challenge

Simplify:

$$\frac{\dfrac{x^2 - 9}{x}}{\dfrac{x^2 + 5x + 6}{x^2}}$$

Challenge

Dividing polynomials can include multiple variables. Divide:

$$\frac{2x^2 + 3xy + y^2}{x + y}$$

- -

Challenge

Write a multi-step absolute-value equation that has no solution.

Saxon Algebra 1

Challenge
Combining Rational Expressions with Unlike Denominators

Challenge

Find the LCD of $\dfrac{3}{mz^2 - mz} - \dfrac{6}{pz^2 - p^2z}$.

- -

Challenge
Graphing Quadratic Functions

Challenge

The height in feet of a dolphin as it jumps out of the water at an aquarium show can be modeled by the function $f(x) = -16x^2 + 32x$, where x is the time in seconds after it exits the water. Find how long the dolphin is in the air.

Saxon Algebra 1

Challenge 97
Graphing Linear Inequalities

Challenge

Find the solution set to the system of inequalities by graphing. The solution set is the region where the areas overlap.

$$y < 3x - 5$$
$$2x - 3y \leq 4$$

- -

Challenge 98
Solving Quadratic Equations by Factoring

Challenge

Factor and solve each quadratic equation. What do they have in common?

$$x^2 + 12x + 36 = 0$$

$$16 = -24x - 9x^2$$

$$16x^2 + 48x = -48x - 144$$

$$81x^2 - 162x + 324 = 162x$$

$$-54x = -9x^2 - 81$$

49

Name _____ Date _____ Class _____

Challenge

Simplify each expression.

1. $\dfrac{\dfrac{15}{m^2} - \dfrac{2}{m} - 1}{\dfrac{4}{m^2} - \dfrac{5}{m} + 4}$

2. $\dfrac{1 - \dfrac{12}{3b + 10}}{b - \dfrac{8}{3b + 10}}$

- -

Name _____ Date _____ Class _____

Challenge

Graph to find the ordered pair that is a solution to all the equations.

$2x^2 + 3x - 5 = 0$

$3x - 2y = 3$

$4x^2 + 3x - 4 = 3$

Saxon Algebra 1

Name _____ Date _____ Class _____

Challenge

Write an inequality that satisfies $x < 0$ OR $x > 0$ and that does not use absolute value.

- -

Name _____ Date _____ Class _____

Challenge

The area of a square is 225 square inches. Write an equation that can be used to find the dimensions of the square. Find the dimensions.

Saxon Algebra 1

Name _____ Date _____ Class _____

Challenge

Simplify these expressions by making the radicand a perfect cube.

$$\frac{4}{\sqrt[3]{2}}$$

$$\frac{1}{\sqrt[3]{3}}$$

- -

Name _____ Date _____ Class _____

Challenge

Complete the square to solve the quadratic equation
$x^2 + 3x = 4$. Then check the answer using factoring.

Saxon Algebra 1

Challenge

The first three stages of Sierpinski's Triangle are shown below.

Stage 1   Stage 2   Stage 3

If the first triangle has an area of 1 square unit, what is the area of the shaded part of the second triangle? the third triangle? What is the area of the shaded part of a triangle at Stage$_n$?

- -

Challenge

Solve $\sqrt{(x + 1)(x + 4)} = \sqrt{(x - 2)(x + 8)}$.

Challenge 107
Graphing Absolute-Value Functions

Challenge

Graph the function
$f(x) = |x^2 - 2x - 4|$ using a table of values.

Describe how this graph differs from the graph of the function
$f(x) = x^2 - 2x - 4$.

- -

Challenge 108
Identifying and Graphing Exponential Functions

Challenge

Research to find some common exponential functions, such as half-life and compound interest. Then choose one function to create and solve your own problem. For example, the half-life of radioactive cobalt is 30 years. After 90 years, how much of 100 grams of cobalt remain?

 Saxon Algebra 1

Challenge 109
Graphing Systems of Linear Inequalities

Challenge

Create a system of inequalities whose solution is (1, 1).

- -

Challenge 110
Using the Quadratic Formula

Challenge

Write your own quadratic equation. It should have two real solutions. Then solve the equation using the quadratic formula.

Challenge 111

Solving Problems Involving Permutations

Challenge

Ann, Bob, Charlie, and Denise are running for President of the Movie Club. The President will get to pick his/her Vice President. However, Ann and Bob don't know each other and won't choose the other. Charlie only wants to work with Ann or Denise. How many ways can the President and Vice President be chosen? Use a tree diagram to solve.

- -

Challenge 112

Graphing and Solving Systems of Linear and Quadratic Equations

Challenge

A system of equations can consist of two quadratic equations. The solution(s) to these systems will still be at the points where the graphs intersect. Solve the system.

$y = x^2 + 2$
$y = -x^2 + 4$

Challenge

Mathematicians define the square root of a negative number to be "imaginary". The letter i is used to denote the square root of -1. So, $\sqrt{-1} = 1i$. For example, $\sqrt{-4} = 2i$, $\sqrt{-9} = 3i$, and so forth. Use the discriminant to find the number of real number solutions of the quadratic equation $x^2 + 1 = 0$. If there are no real number solutions, then find the imaginary solutions using $\sqrt{-1} = 1i$.

- -

Challenge

Use a graphing calculator to graph these functions:

$f(x) = \sqrt[3]{x}$

$f(x) = \sqrt[4]{x}$

$f(x) = \sqrt[5]{x}$

$f(x) = \sqrt[6]{x}$

Is there a pattern to the graphs? Explain your answer.

Challenge **115**
Graphing Cubic Functions

Challenge

Describe the end behavior of each graph:
   linear equation

   quadratic equation

   cubic equation

Use this knowledge to predict the end behavior of the graph of $y = x^4$.
Then graph $y = x^4$ to check your predicton.

— —

Challenge **116**
Solving Simple and Compound Interest Problems

Challenge

Antwaan deposited $5000 in a savings account that paid 5.75% interest
compounded semi-annually. After 6 years, he withdrew $2500. What is the
total amount in the account 13 years after the account was opened?

       Saxon Algebra 1

Name _____ Date _____ Class _____

Challenge

Identify trigonometric ratios that have the value $\frac{5}{12}$.

- -

Name _____ Date _____ Class _____

Challenge

There are 13 different toppings available at a pizza place. A family chooses 3 different toppings. What is the probability that the toppings are three of the following: pineapple, mushrooms, olives, tomatoes, or peppers?

Saxon Algebra 1

Challenge

Describe a situation that can be modeled by an exponential function.

- -

Challenge

A rectangular garden has a length twice as long as its width. In the middle of the garden is a triangular fountain with height and base equal to the garden's width. What is the probability that a bird would land on the fountain?

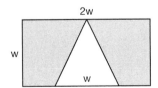

Name _____ Date _____ Class _____

The irrational number pi is a non-repeating, non-terminating decimal number, 3.14... .

Evaluate each power below and place the last digit of each answer in the blanks provided to discover the first 20 decimal places of pi.

1. 9^2 _____

2. 2^6 _____

3. $3^2 + 2^5$ _____

4. 25^2 _____

5. 7^2 _____

6. 2^9 _____

7. 4^4 _____

8. $10^2 - 5^2$ _____

9. 7^3 _____

10. 5^4 _____

11. 2^7 _____

12. 9^3 _____

13. $8^3 - 5^3$ _____

14. $13^2 \cdot 8^0$ _____

15. $11^2 + 8^3$ _____

16. $4^2 + 6^2$ _____

17. 3^5 _____

18. $2^2 \cdot 2^5$ _____

19. $3^3 \cdot 2^5$ _____

20. $12^2 - 2^7$ _____

3. __ . __

Saxon Algebra 1

Simplify each expression.

1. $\dfrac{\sqrt{\dfrac{6 + 16 \cdot 2^3 - 12 + 22}{2(25 - 16) - 14}}}{3^3 \div 9 + 3}$

        _____

2. $\dfrac{\left|-200 - (-450) - 850\right| \times 2}{2^5 + 2^3} - \left[5(-16 - 4) + \dfrac{18 + 2}{2^2}\right]$

        _____

3. $\dfrac{(3^2 + 5 \cdot 4^2 - 12 \cdot 5^2 + 20) + 20(10) + 1}{2 + 3}$

        _____

4. $\dfrac{\dfrac{9 - 6 \cdot 2 - 4 \cdot 6}{-3}}{\sqrt{\dfrac{2^3 \cdot 2^2 \cdot 2^1}{(\sqrt{169} - 5) \cdot (5 + 3)}}}$

        _____

5. $16 \cdot \dfrac{\sqrt{150 - 120 - 45 + 40}}{\sqrt{225} - \dfrac{150}{-2} + 6 \cdot 10}$

        _____

6. $\left|\dfrac{\sqrt{4^2 \cdot 4^2 \cdot 4^2} - 120 \cdot \dfrac{2}{5}}{-2}\right|$

        _____

 Saxon Algebra

Enrichment 7
Digits

Using the digits 1, 2, 3 and 4 and addition, subtraction, multiplication, division, parentheses, and exponents, write an expression equivalent to the numbers 1 to 20.

- You must use all four digits in each expression.

- You may use any of the operations but each symbol may be used only once in each expression.

An example has been done for you. There may be more than one correct expression for a given number.

$1 = 3 \cdot 2 - (4 + 1)$ $2 = $ _____

$3 = $ _____ $4 = $ _____

$5 = $ _____ $6 = 3^1 \cdot (4 - 2)$

$7 = $ _____ $8 = $ _____

$9 = $ _____ $10 = $ _____

$11 = $ _____ $12 = $ _____

$13 = $ _____ $14 = $ _____

$15 = $ _____ $16 = $ _____

$17 = $ _____ $18 = $ _____

$19 = $ _____ $20 = $ _____

Name _____ Date _____ Class _____

Evaluate each expression in Column 1 for the value given and match it to the correct answer in Column 2. Then, write the letter above the corresponding exercise number to discover a common mathematical property.

Column 1   **Column 2**

1. $x - 9$, for $x = 3$ **B** -8

2. $2m + 3$, for $m = -2$ **D** 8

3. $\dfrac{t}{2} + 6$, for $t = 10$ **E** 11

4. $15 - 4y$, for $y = 3$ **I** -1

5. $-3x + 1$, for $x = -3$ **O** -7

6. $\dfrac{5}{2} - \dfrac{d}{2}$, for $d = 3$ **P** -11

7. $\dfrac{p}{3} + 12$, for $p = -9$ **R** 3

8. $-6 + 4x$, for $x = -\dfrac{1}{2}$ **S** 1

9. $\dfrac{a}{3} - 10$, for $a = 9$ **T** -6

10. $7f + 8$, for $f = 0$ **U** 10

11. $-5x - 6$, for $x = 1$ **V** 0

12. $-\dfrac{c}{5} + 3$, for $c = 15$ **Y** 9

___ ___ ___ ___ ___ ___ ___ ___ ___ ___ ___ ___
10   2   6   1   4   2   8   5   1   2   12   3

___ ___ ___ ___ ___ ___ ___ ___
11   4   9   11   3   4   1   7

Saxon Algebra 1

Complete the crossword puzzle.

ACROSS

1. $\frac{5}{9} + \frac{13}{9} =$ _____

2. $4.8 \div 0.16 =$ _____

6. $\frac{2}{5} \div \frac{2}{7} =$ seven _____

7. $1.09 + 4.2 =$ five and twenty-nine _____

9. $\frac{1}{10} + \frac{3}{5} =$ _____ tenths

10. $20 \div 0.5 =$ _____

13. $10.29 - 7.09 =$ three and two _____

16. $\frac{1}{3} \div \frac{4}{3} =$ one _____

17. $14 \times 22 =$ three hundred and _____

DOWN

1. $\frac{1}{2} - \frac{1}{6} =$ one _____

2. $1\frac{3}{8} \times 1\frac{3}{5} =$ two and _____ hundredths

3. $12 - 11.999 =$ one _____

4. $1.06 + 11.94 =$ _____

5. $180 \div 12 =$ _____

8. $\frac{1}{2} \times \frac{6}{7} =$ three _____

10. $46 + 4 =$ _____

11. $\frac{2}{5} \times \frac{1}{4} =$ one _____

12. $\frac{7}{8} - \frac{3}{8} =$ one _____

14. $0.2 \times 1.5 =$ _____ tenths

15. $158 - 98 =$ _____

Saxon Algebra 1

Enrichment
Good Luck Squares

13

Shade each box that contains a perfect square number. Then, find its positive square root and shade that box as well.

50	17	31	80	42	99	69	300	18	115	46	91	63
72	94	65	89	550	10	8	121	97	950	59	150	73
39	615	54	78	16	106	32	112	15	88	19	825	29
125	60	215	211	225	117	377	76	9	82	105	52	85
22	800	81	1	12	250	21	47	196	25	13	115	45
815	4	114	37	116	6	53	7	500	101	525	36	68
51	400	43	113	93	58	40	325	650	715	71	11	28
750	49	725	23	111	144	925	64	74	33	315	169	61
86	98	14	625	900	107	256	109	30	100	5	700	56
57	75	83	110	102	415	20	90	915	95	350	108	77
600	70	66	24	450	48	25	425	850	26	104	84	38
34	87	44	92	79	515	35	103	67	96	55	41	27

Saxon Algebra 1

Name _____ Date _____ Class _____

Enrichment **17**
Number Puzzles

Given the following scenarios, fill in the blank with one of these numbers.

6.5	500	-4	1	-75	$\sqrt{17}$	-5	18	$\frac{11}{9}$
$\frac{5}{6}$	$\sqrt{15}$	$\frac{96}{4}$	4.5	$-\sqrt{11}$	0	$\frac{2}{3}$	300	-12

1. When you square my value and subtract it from 23, the result is the square root of 36. I am an irrational number. What is my number? _____

2. If you cube my value and divide it by the quotient of 150 and 6, the result is my value. I am not a natural number, but I am a rational number. What is my number? _____

3. When you take my value and multiply it by -8, the result is an integer greater than -220. If you take the result and divide it by the sum of -10 and 2, the result is my value. I am a rational number. What is my number? _____

4. If you take my absolute value and subtract the square of 16, the result is a double-digit number whose prime factorization is $11 \times 2 \times 2$. I am a natural number. What is my number? _____

5. When you add 5 to my value and subtract $1\frac{1}{2}$, the result is twice the square root of 25. I am a terminating decimal. What is my number? _____

6. If you take 3 to the value of my power, the result is a non-terminating, non-repeating decimal. If you take 3 and raise it to the power of the absolute value of my value, the result is 81. I am an integer. What is my number? _____

7. When you divide my value by the absolute value of -6, the result is a fraction that when simplified has a numerator that is the first natural number and a denominator that is a perfect square. If you take the square root of the reciprocal of this fraction, the result is 3. I am a real number. What is my number? _____

© Saxon. All rights reserved. 67 **Saxon** Algebra 1

Name _____ Date _____ Class _____

Write the letter that corresponds to the coordinate pair in the blank provided to discover a common mathematical expression.

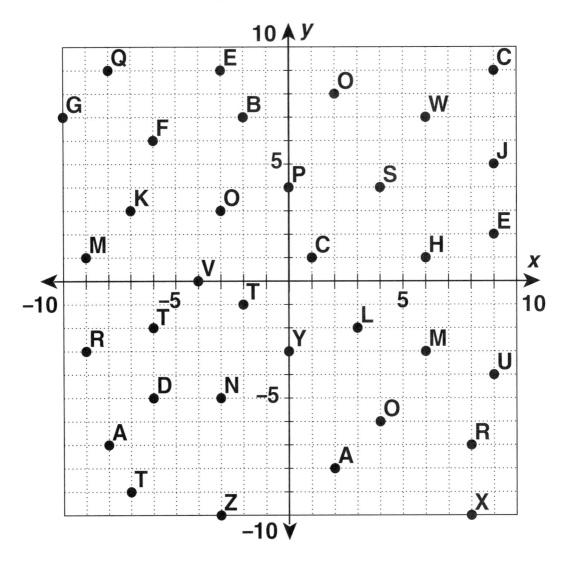

$\overline{}$ $\overline{}$ $\overline{}$ $\overline{}$ $\overline{}$ $\overline{}$ $\overline{}$ $\overline{}$
(−10, 7) (8, −7) (9, 2) (2, −8) (−6, −2) (−3, 9) (4, 4) (−2, −1)

$\overline{}$ $\overline{}$ $\overline{}$ $\overline{}$ $\overline{}$ $\overline{}$
(1, 1) (4, −6) (−9, 1) (6, −3) (−3, 3) (−3, −5)

$\overline{}$ $\overline{}$ $\overline{}$ $\overline{}$ $\overline{}$ $\overline{}$
(−6, 6) (−8, −7) (9, 9) (−7, −9) (2, 8) (−9, −3)

Saxon Algebra 1

Enrichment 23
Magic Square

The rows, columns, and diagonals of a magic square all have the same sum.
Create a magic square by solving these equations.

1.	2.	3.	4.
5.	6.	7.	8.
9.	10.	11.	12.
13.	14.	15.	16.

1. $\dfrac{x}{2} = 4$

2. $t + 7 = 2$

3. $-2m = 12$

4. $k - 10 = -5$

5. $-b = 3$

6. $1 = \dfrac{y}{2}$

7. $-9h = -27$

8. $n + 4 = 4$

9. $-12 = q - 13$

10. $16 = -8x$

11. $-5 = f - 4$

12. $\dfrac{t}{2} = 2$

13. $r = 6 - 10$

14. $y - 1 = 6$

15. $-3 = -\dfrac{t}{2}$

16. $-3m = 21$

What is the magic sum? _____

Saxon Algebra 1

Name _____ Date _____ Class _____

A relation is a function if for every number in the domain, there is one and only one number in the range. Another way to describe a function is to determine if it is onto. A function is onto if and only if all the numbers of the range are paired with all of the numbers in the domain.

For example, the function in Table 1 is described as onto because each number in the range has a number from the domain assigned to it.

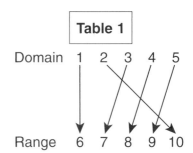

The function in the Table 2 is not onto because the number 7 in the range does not have a number in the domain assigned to it.

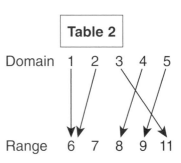

Determine if each function can be described as onto.

1. _____

2. _____

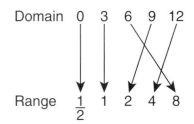

3. _____

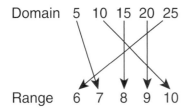

4. _____

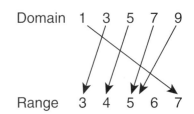

Saxon Algebra 1

Name _____ Date _____ Class _____

A perfect number is a number which is the sum of its own positive factors (other than itself). For example, the following numbers are perfect.

$6 = 1 + 2 + 3$

$28 = 1 + 2 + 4 + 7 + 14$

$496 = 1 + 2 + 4 + 8 + 16 + 31 + 62 + 124 + 248$

The next perfect number is 8128. What is the fifth perfect number? _____

To find the answer, solve each equation and write your answer in the blanks provided at the bottom of the page.

1. $7x - 12 = 9$

2. $\dfrac{x}{3} - 4 = -3$

3. $25 - 3x = 10$

4. $-2 = \dfrac{x}{5} - 3$

5. $4(2x - 1) = -4$

6. $7(2 - x) = -7$

7. $25x - 10x = 40 + 5$

8. $15 - \dfrac{x}{6} = 14$

$$\overline{}\ \overline{}\ ,\ \overline{}\ \overline{}\ \overline{}\ ,\ \overline{}\ \overline{}\ \overline{}$$
  1.   2.   3.   4.   5.   6.   7.   8.

Saxon Algebra 1

Name _____ Date _____ Class _____

Diophantus (about 200–284) is known to some as the 'father of algebra'. He studied primarily the solutions of algebraic equations and the theory of numbers.

One type of equation he studied has the form $ax + by = c$ where a, b, and c are all integers and the solutions to the equation (x, y) are also integers. These types of equations are now known as Diophantine Equations. They can be quite difficult to solve and many times the only way to solve them is by guessing and checking.

Solve each Diophantine Equation. Find at least one pair of positive integers for x and y that make the equation true.

1. $3x + 4y = 12$

    a. Solve the equation for y. _____

    b. What number must x be divisible by?

       Why? _____

    c. Find at least one solution. _____

2. $-2x + 3y = -9$

    a. Solve the equation for y. _____

    b. What number must x be divisible by?

       Why? _____

    c. Find at least one solution. _____

3. $x - 2y = 10$

4. $-4x + y = 15$

5. $8x - 19y = 100$

6. $3x + 7y = 35$

7. $-5x + 11y = 30$

8. $3x + 4y = 32$

9. $7x - y = 14$

10. $-3x + 5y = 9$

Saxon Algebra 1

Name _____ Date _____ Class _____

An important concept in business is the ability to make a profit. Profit is equal to the amount of sales minus the cost of production. If the sales are greater than the cost, the business makes a profit. If the sales are less than the cost, the business is losing money.

Use the information below to answer each question.

A manufacturer of compact-disc players sells them to a retailer for $45 each. It costs the manufacturer $200 plus $25 each to produce the compact-disc player.

1. Write a function, *s*, to represent the total amount of sales of

   compact-disc players, *n*. _____

2. Write a function, *c*, to represent the total cost of producing the

   compact-disc players, *n*. _____

3. Graph the functions *s* and *c* on the same coordinate grid.

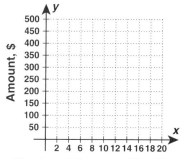

Number of Compact-Disc Players

4. For what dollar amount is the sales and the cost equal? _____

5. For what value of *n* is the sales and the cost equal? _____

6. Write an inequality that represents the value(s) of *n* for which the cost is more than the sales.

   _____

7. Write an inequality that represents the value(s) of *n* for which the manufacturer makes a profit.

   _____

       Saxon Algebra 1

Enrichment **31**
Proportions in Paint

Thousands of different paint colors are possible because *colorants*, or dyes, are used in different proportions. For example, one type of beige requires 2 parts black, 1 part maroon, and 15 parts deep gold. The ratio describing this situation is shown below.

$$\text{black:maroon:deep gold} = 2:1:15$$

Suppose 5 ounces of black colorant are used to make the beige. Then how many ounces of maroon and deep gold colorant need to be used?

Step 1 Find the amount of maroon colorant.

$\dfrac{\text{black} \rightarrow 2}{\text{maroon} \rightarrow 1}$ Write a ratio comparing black to maroon.

$2/1 = 5/x$ Write a proportion. Let x be the ounces of maroon.

$2(x) = 1(5)$ Use cross products.

$2x = 5$ Simplify.

$2x/2 = 5/2$ Divide both sides by 2.

$x = 2.5$ So, 2.5 ounces of maroon colorant are used.

Step 2 Find the amount of deep gold colorant.

$\dfrac{\text{maroon} \rightarrow 1}{\text{deep gold} \rightarrow 15}$ Write a ratio comparing maroon to deep gold.

$1/15 = 2.5/y$ Write a proportion. Let y be the ounces of deep gold.

$1(y) = 15(2.5)$ Use cross products.

$y = 37.5$ So, 37.5 ounces of deep gold colorant are used.

1. An almond color is made by using 4 parts of new green, 3 parts of maroon, and 12 parts of deep gold.

   a) Write a ratio comparing the amounts of new green, maroon, and deep gold. _____

   b) If 14 ounces of new green colorant are used, how much maroon and deep gold colorants are needed? _____

2. Periwinkle is made by mixing thalo green, thalo blue, and magenta in the ratio 16:20:35. If 15 ounces of thalo blue colorant are used, how much thalo green and magenta colorants need to be used?

   _____

3. Navy blue is made by using the following colorants: 55 parts black, 14 parts blue, and 50 parts magenta. If 180 grams of magenta colorant are used, how much black and blue colorants are needed?

   _____

 Saxon Algebra

Name _____ Date _____ Class _____

What did one math book say to the other math book? _____

To discover the answer, write the equivalent fraction and decimal for each percent. Then, cross out each answer in the code box. The remaining letters will reveal the solution.

1. 15% _____ _____   **2.** 20% _____ _____

3. 32% _____ _____   **4.** 50% _____ _____

5. 60% _____ _____   **6.** 85% _____ _____

7. 12% _____ _____   **8.** 5% _____ _____

9. 72% _____ _____   **10.** 25% _____ _____

11. 40% _____ _____   **12.** 95% _____ _____

13. 70% _____ _____   **14.** 75% _____ _____

15. 44% _____ _____

$\frac{8}{25}$	$\frac{3}{4}$	$\frac{3}{20}$	$\frac{3}{50}$	$\frac{19}{20}$	$\frac{11}{25}$	$\frac{4}{5}$	$\frac{1}{2}$	$\frac{17}{20}$	$\frac{1}{10}$	$\frac{3}{25}$	$\frac{7}{20}$
P	E	T	B	A	S	O	N	H	Y	E	D
$\frac{13}{20}$	$\frac{18}{25}$	$\frac{9}{10}$	$\frac{1}{4}$	$\frac{2}{5}$	$\frac{11}{50}$	$\frac{3}{10}$	$\frac{3}{5}$	$\frac{7}{10}$	$\frac{11}{20}$	$\frac{1}{20}$	$\frac{1}{5}$
O	N	I	S	E	H	A	T	I	V	W	A
0.2	0.1	0.6	0.05	0.8	0.5	0.9	0.15	0.55	0.7	0.4	0.65
L	E	B	T	P	O	R	E	O	S	H	B
0.95	0.35	0.75	0.32	0.44	0.13	0.85	0.72	0.12	0.73	0.3	0.25
D	L	P	A	S	E	T	H	R	M	S	E

Saxon Algebra 1

Name _____ Date _____ Class _____

Look at the graph of the equation $\frac{1}{2}x + \frac{1}{3}y = 1$ and
$\frac{1}{-4}x + \frac{1}{5}y = 1$ shown on the right. The *y*-intercept for the
first equation is 3, so you know that the point (0, 3) lies on
the line. Another point containing 0 also lies on the line; it is
the point (2, 0). The *x*-coordinate of the point at which the line
crosses the *x*-axis is called the *x*-intercept.

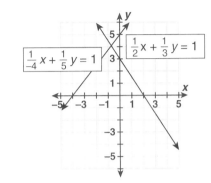

You can see that for the second equation, the
y-intercept is 5.

What is the *x*-intercept of the second equation? _____

An equation in the form $\frac{1}{a}x + \frac{1}{b}y = 1$ is in intercept-intercept form. In this

form, *a* is the *x*-intercept and *b* is the *y*-intercept.

Look back at the equation $\frac{1}{-4}x + \frac{1}{5}y = 1$. Since it is in
intercept-intercept form, you know the following relationships.

$\frac{1}{a} = \frac{1}{-4}$ $\frac{1}{b} = \frac{1}{5}$ How would you determine *a* and *b*?

$a = -4$ $b = 5$ _____

You can use the intercept-intercept for to determine the slope of the line.
Recall that slope equals rise over run. You can use the two intercepts to
count the rise and run on the graph. Applying the definition of slope, you
can calculate the slope using the intercepts *a* and *b*.

$$m = \frac{y_2 - y_1}{x_2 - x_1} = \frac{0 - b}{a - 0} = -\frac{b}{a}$$

What is the slope of the line described by the equation $\frac{1}{2}x + \frac{1}{3}y = 1$?

$m = $
____

**For each equation, determine the *x*-intercept, the *y*-intercept and the
slope of the line.**

1. $x + \frac{1}{6}y = 1$ **2.** $x - y = 1$ **3.** $x + y = 1$

_____   _____   _____

4. $\frac{1}{2}x + \frac{1}{6}y = 1$ **5.** $x + y = 2$ **6.** $x - 2y = 2$

_____   _____   _____

Saxon Algebra 1

Name _____ Date _____ Class _____

The Triangle Inequality Theorem states that "For any triangle, the sum of the lengths of any two sides is greater than the length of the third side". This inequality defines the existence of a triangle. There is a theorem in geometry that determines whether a given triangle is a right triangle, obtuse triangle, or acute triangle.

A right triangle has exactly one 90 degree angle.

An obtuse triangle has exactly one angle greater than 90 degrees.

An acute triangle has no angle with a measure greater than or equal to 90 degrees.

In a triangle with sides *a, b,* and *c* with *c* being the longest side:

If $c^2 > a^2 + b^2$, the triangle is obtuse.

If $c^2 = a^2 + b^2$, the triangle is a right triangle.

If $c^2 < a^2 + b^2$, the triangle is acute.

Determine whether the triangles, with these given side lengths, are acute, right, or obtuse.

1. 2, 3, 4

_____

2. 3, 4, 5

_____

3. 6, 6, 7

_____

4. 7, 20, 24

_____

5.

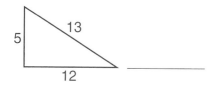

_____

6.

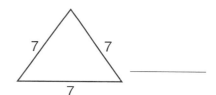

_____

7.

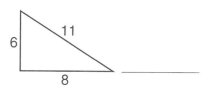

_____

8.

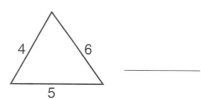

_____

9. The longest side of an acute triangle measures 12 inches. One of the shorter sides is 7 inches. Express the length of the third side as an inequality.

_____

10. The two shorter sides of an obtuse triangle measure 5 cm and 10 cm. Express the length of the third side as an inequality.

_____

Saxon Algebra 1

Name _____ Date _____ Class _____

Enrichment 58
A Maze of Multiplication

Travel around the maze as you solve each multiplication problem. Each answer will touch the previous answer. Shade your answers in the grid to discover an amazing scientific accomplishment. Answers may be written up, down, forwards, backwards, or diagonally and boxes may be used more than once if needed.

1. $x(x + 3)$

2. $(x + 1)(x + 7)$

3. $(x - 3)(x + 5)$

4. $x(2x - 8)$

5. $8x(1 - x)$

6. $(x + 3)(x - 7)$

7. $(2x + 1)(x - 6)$

8. $3x(9 - x)$

9. $(5x - 1)(3x + 1)$

10. $(4x + 2)(3x + 3)$

11. $(x - 8)(x + 8)$

12. $5x(9 - 4x)$

13. $2(7x - 3)$

14. $(x - 10)(x - 3)$

15. $(5 + 2x)(6 - x)$

16. $(3x - 7)(3x + 7)$

17. $2x(5 - 2x)$

18. $(8x + 1)(5x - 6)$

19. $(3x - 2)(3x + 2)$

20. $(7x + 3)(x - 5)$

Finish here. ←⌐   ⌐→ Start here.

x^2	−	$7x$	x^2	11	+	$7x$	−	8	15	x^2	+	$3x$	+	2	x^2	+	4	+	8	−	x^2	+	$7x$	$3x$	x^2
8	11	$9x$	$4x^2$	x	$9x$	14	−	19	$4x^2$	−	$5x$	+	x^2	x	+	$4x^2$	8	x^2	+	$9x$		11	23	+	14
$9x^2$	−	49	$10x$	8	+	x^2	4	23	11	+	$32x$	$3x$	23	+	8	+	$3x$	+	4	+	4	x	+	x^2	4
$7x$	$2x^2$	x	$3x$	−	11	$3x$	+	14	x^2	4	$5x$	−	$7x$	+	$8x$	14	+	x^2	+	$7x$	+	$5x$	$4x^2$	−	$5x$
4	+	−	$4x^2$	+	$4x^2$	19	+	$7x$	$9x$	14	$3x$	8	$7x^2$	x	−	+	−	−	−	$5x$	+	8	19	x	$7x$
−	$7x$	8	$7x$	−	$9x$	$40x^2$	−	$43x$	−	6	$9x^2$	−	4	+	$7x$	−	7	x^2	+	$2x$	−	15	$2x^2$	+	$5x$
x	+	19	−	+	−	23	+	$9x$	$3x$	11	+	19	11	4	+	4	11	−	−	$5x$	+	11	$3x$	−	$9x$
x^2	−	x^2	$3x$	−	30	$4x^2$	8	x^2	$4x^2$	4	$3x$	x^2	$9x$	+	x^2	+	8	x^2	x	14	8	+	x^2	x	$8x$
+	$7x$	4	−	+	−	$5x$	x	+	$7x$	+	11	+	$4x^2$	8	+	14	x	23	4	+	19	14	11	−	$4x^2$
$3x$	4	8	$13x$	−	+	64	−	x^2	6	+	$18x$	+	$12x^2$	4	$4x^2$	$3x$	$2x^2$	21	−	$4x$	−	x^2	$8x^2$	+	4
−	x	−	11	+	$45x$	$7x$	+	x	11	$7x$	8	4	1	+	$9x$	−	$7x$	$4x^2$	+	8	$5x$	+	8	+	11
x	x^2	$3x$	$9x$	−	4	19	4	+	$3x$	+	14	−	8	8	$11x$	+	$9x$	+	+	$7x$	23	x^2	+	x	$5x$
6	−	$14x$	$20x^2$	+	14	x^2	−	11	x^2	−	$2x$	8	$3x$	−	$7x$	8	+	$3x$	4	x^2	+	14	$9x$	$3x$	x^2
+	$7x$	11	$7x$	x	$9x$	$4x^2$	8	x^2	−	+	11	$7x$	6	+	x	x^2	$4x^2$	$5x$	+	$5x$	$9x$	11	$7x$	+	8
x^2	+	8	x^2	+	−	$3x$	−	$5x$	$15x^2$	$3x^2$	−	$27x$	$9x$	−	−	19	4	x^2	$5x$	−	$7x$	−	x	x^2	−

Saxon Algebra 1

Name _____ Date _____ Class _____

Pascal's Triangle is a geometric arrangement of numbers. These numbers represent the binomial coefficients. That is, they represent the coefficients of the terms of the expansion of $(x + y)^n$. The first seven rows of Pascal's Triangle look like this.

Row 0   1

Row 1   1   1

Row 2   1   2   1

Row 3   1   3   3   1

Row 4   1   4   6   4   1

Row 5   1   5   10   10   5   1

Row 6   1   6   15   20   15   6   1

Notice that each number is the sum of the two numbers above it.

For example what two numbers were added to get 10 in the 5th row? _____

What are the numbers for the 7th row? _____

As an example, find $(x + y)^3$.

Look at row 3, what are the coefficients of the expansion? _____

The first term of the expansion starts with the highest power of x, namely x^3, and the lowest power of y, namely $y^0 = 1$. The power of x increases by 1 for each successive term and the power of y increases by 1 for each successive term.

$$(x + y)^3 = \underline{1 \cdot x^3 \cdot y^0} + \underline{3 \cdot x^2 \cdot y^1} + \underline{3 \cdot x^1 \cdot y^2} + \underline{1 \cdot x^0 \cdot y^3}$$

$$= \quad x^3 \quad + \quad 3x^2y \quad + \quad 3xy^2 \quad + \quad y^3$$

Expand each of the following polynomials.

1. $(x + y)^4$

2. $(x + 1)^5$

_____ _____

3. $(x + 3)^4$

4. $(x + 2)^3$

_____ _____

5. $(x + 1)^9$

6. $(x + 2y)^5$

_____ _____

_____ _____

Saxon Algebra 1

Name _____ Date _____ Class _____

Two types of variation have been discussed thus far. Direct variation is an equation of the form $y = kx$ while inverse variation is an equation of the form $y = \frac{k}{x}$ or $xy = k$.

Two important laws from Chemistry arise from the idea of variation. Boyle's Law states that the volume of a gas at a given temperature varies inversely with applied pressure. Mathematically, this inverse variation can be expressed as $V = \frac{k}{P}$, where V is the volume, P is the applied pressure, and k is a constant. Charles' Law states that the volume of a gas at a given pressure varies directly with temperature. Mathematically, this direct variation can be stated $V = kT$, where V is the volume, T is the temperature, and k is a constant.

Combining the two laws results in what is called a joint variation: The volume of a gas varies directly with the temperature and inversely with pressure.

Answer each question.

1. Write the mathematical statement for the joint variation of the two gas laws.

_____

2. a. If the volume of a sample of gas is 3.241 L under a pressure of 0.20 atm at a temperature of 300 Kelvin, find k.

_____

  b. If the pressure was adjusted to 0.50 atm and the temperature was changed to 320 Kelvin, determine the volume of the sample of gas.

_____

  c. If the temperature of the sample was held constant at 320 Kelvin, what would the pressure need to be adjusted to in order to return the volume to 2 L?

_____

3. a. Suppose a 5 L sample of gas under went the following changes: the pressure was changed from 0.1 atm to 0.07 atm, and the temperature was changed from 400 Kelvin to 320 Kelvin. Determine the volume of the sample of gas.

_____

  b. Holding the pressure constant at 0.07 atm, what change in temperature would return the gas to a volume of 5 L?

_____

Saxon Algebra

Name _____ Date _____ Class _____

A matrix is a rectangular array of numbers enclosed in a single set of brackets. If each equation in a system of equations is written in standard form, you can represent the system with a matrix equation. The matrix equation is made up of three matrices; one for the coefficients on the variables x and y, one for the variables x and y, and one for the constants.

For example, the system of equations $\begin{cases} 3x - y = 6 \\ x + y = -2 \end{cases}$ is represented by the matrix

equation $\begin{bmatrix} 3 & -1 \\ 1 & 1 \end{bmatrix}\begin{bmatrix} x \\ y \end{bmatrix} = \begin{bmatrix} 6 \\ -2 \end{bmatrix}$.

Determine which system of equations represents the correct matrix equation.

1. _____ $\begin{cases} x + y = 8 \\ x - y = 2 \end{cases}$
   a. $\begin{bmatrix} 3 & -1 \\ 6 & 2 \end{bmatrix}\begin{bmatrix} x \\ y \end{bmatrix} = \begin{bmatrix} 4 \\ -8 \end{bmatrix}$

2. _____ $\begin{cases} 3x - y = 4 \\ 6x + 2y = -8 \end{cases}$
   b. $\begin{bmatrix} 1 & -1 \\ 2 & 3 \end{bmatrix}\begin{bmatrix} x \\ y \end{bmatrix} = \begin{bmatrix} 2 \\ 9 \end{bmatrix}$

3. _____ $\begin{cases} x - y = 2 \\ 2x + 3y = 9 \end{cases}$
   c. $\begin{bmatrix} -5 & 8 \\ 10 & 3 \end{bmatrix}\begin{bmatrix} x \\ y \end{bmatrix} = \begin{bmatrix} 21 \\ 15 \end{bmatrix}$

4. _____ $\begin{cases} -5x + 8y = 21 \\ 10x + 3y = 15 \end{cases}$
   d. $\begin{bmatrix} 1 & 1 \\ 1 & -1 \end{bmatrix}\begin{bmatrix} x \\ y \end{bmatrix} = \begin{bmatrix} 8 \\ 2 \end{bmatrix}$

Create a matrix equation for each system of equations.

5. $\begin{cases} x - 5y = 0 \\ 2x - 3y = 7 \end{cases}$
6. $\begin{cases} 4x + 3y = 19 \\ 3x - 4y = 8 \end{cases}$
7. $\begin{cases} 5x + 3y = 12 \\ 4x - 5y = 17 \end{cases}$

_____   _____   _____

8. $\begin{cases} x + y = 7 \\ x - y = 9 \end{cases}$
9. $\begin{cases} 12x - 9y = 114 \\ 12x + 7y = 82 \end{cases}$
10. $\begin{cases} 2x - 3y = -4 \\ x + 3y = 7 \end{cases}$

_____   _____   _____

11. $\begin{cases} \frac{1}{2}x + y = 12 \\ x + \frac{1}{5}y = 10 \end{cases}$
12. $\begin{cases} \frac{2}{3}x + \frac{1}{3}y = -9 \\ \frac{1}{4}x + \frac{3}{4}y = 16 \end{cases}$
13. $\begin{cases} 1.2x - 1.6y = 2.4 \\ -0.8x + 0.2y = -1.2 \end{cases}$

_____   _____

_____

Saxon Algebra 1

Enrichment **72**
Factoring Fun

What is the Roman numeral MMDCCXLVIII equivalent to in Arabic numerals? _____

To discover the answer, factor each trinomial. Then, answer the four questions at the bottom of the page. Place each answer in the blank above the corresponding exercise number.

1. $x^2 - 6x + 8$ **2.** $x^2 + x - 6$ **3.** $x^2 - 5x + 4$

4. $x^2 + x - 2$ **5.** $x^2 - 2x - 8$ **6.** $x^2 + x - 12$

7. $x^2 - x - 12$ **8.** $x^2 - 3x - 4$ **9.** $x^2 - 3x + 2$

10. $x^2 - 8x + 16$ **11.** $x^2 - 2x - 15$ **12.** $x^2 + 5x + 6$

13. $x^2 - 5x + 6$ **14.** $x^2 + x - 20$ **15.** $x^2 - x - 2$

16. $x^2 + 4x - 5$ **17.** $x^2 + 2x - 8$ **18.** $x^2 + 3x - 10$

1. How many of the trinomials have a factor of $x + 1$?

2. How many of the trinomials have a factor of $x - 2$?

3. How many of the trinomials have a factor of $x + 3$?

4. How many factors of $x - 4$ do you see?

_____ _____ _____ _____
   1   2   3   4

                        Saxon Algebra **1**

Enrichment
Fourth Degree Trinomials 75

Sometimes it is possible to write a trinomial of the fourth degree, $a^4 + a^2b^2 + b^4$, as a difference of two squares and then factor.

Example: Factor $4a^4 - 21a^2b^2 + 9b^4$.

Step I Find the square roots of the first and last terms.

$$\sqrt{4a^4} = 2a^2 \qquad\qquad \sqrt{9b^4} = 3b^2$$

Step II Find twice the product of the square roots from the terms in Step 1.

$$2(2a^2)(3b^2) = 12a^2b^2$$

Step III Split the middle term of the trinomial into two parts. One part is either the answer from the Step II or its opposite. The other part should be the opposite of a perfect square.

$$-21a^2b^2 = -12a^2b^2 - 9a^2b^2$$

Step IV Rewrite the trinomial as the difference of two squares and then factor.

$$
\begin{aligned}
4a^4 - 21a^2b^2 + 9b^4 &= (4a^4 - 12a^2b^2 + 9b^4) - 9a^2b^2 \\
&= (2a^2 - 3b^2)^2 - 9a^2b^2 \\
&= [(2a^2 - 3b^2) - 3ab]\,[(2a^2 - 3b^2) - 3ab] \\
&= (2a^2 + 3ab - 3b^2)(2a^2 - 3ab - 3b^2)
\end{aligned}
$$

Factor each trinomial.

1. $16d^4 + 7d^2 + 1$

_____

2. $p^4 + p^2 + 1$

_____

3. $4x^4 - 13x^2 + 1$

_____

4. $4x^4 - 9x^2y^2 + 16y^4$

_____

5. $9r^4 + 26r^2s^2 + 25s^4$

_____

6. $4a^4 - 5a^2c^2 + 25c^4$

_____

Saxon Algebra 1

Name _____ Date _____ Class _____

Measurements made with a ruler are not precise. The greatest possible error in a measurement is one-half of the unit of measure. For example, if you measure a line segment to be 4.7 cm, the greatest possible error is one-half of 0.1 cm, or 0.05 cm.

The minimum length of this line segment is $(4.7 - 0.05)$ or 4.65 and the maximum length is $(4.7 + 0.05)$ or 4.75. Written as an inequality it can be represented as $4.65 \leq \ell < 4.75$, where ℓ is the length.

Find the greatest possible error for each of the following.

1. a line segment, ℓ, 3.2 m long

2. an object, b, weighing 1.5 g

_____

_____

3. a pitcher, p containing 2.75 L

4. a distance, d of 1.75 miles

_____

_____

Determine the inequality representing Exercises 1 to 4.

5. a line segment

6. an object

_____

_____

7. a pitcher

8. a distance

_____

_____

Determine the maximum and minimum areas of each figure and express each as an inequality.

9. A rectangle having dimensions of 12 feet by 8 feet

_____

_____

10. A square having a side length of 10.34 cm

_____

_____

Saxon Algebra 1

Enrichment
Sum and Differences of Cubes
83

While it is possible to factor the sum and differences of two squares, it is also possible to factor the sum and differences of two cubes.

The sum of two cubes can be factored in the following way:

$$a^3 + b^3 = (a + b)(a^2 - ab + b^2)$$

The differences of two cubes can be factored in the following way:

$$a^3 - b^3 = (a - b)(a^2 + ab + b^2)$$

Factor each of the following.

1. $r^3 - s^3$

2. $x^3 + y^3$

_____

_____

3. $x^3 + 8$

4. $n^3 - 64$

_____

_____

5. $8y^3 + 27$

6. $pq^3 - 64p$

_____

_____

Express each of the following as the sum or difference of two cubes.

7. $(m - 1)(m^2 + m + 1)$

8. $(2 + 3t)(4 - 6t + 9t^2)$

_____

_____

9. $(b - 64)(b^2 + 4b + 16)$

10. $(x + 7)(x^2 - 7x + 49)$

_____

_____

11. $(2y - 1)(4y^2 + 2y + 1)$

12. $(3 + 2t)(9 - 6t + 4t^2)$

_____

_____

13. $(s + 10)(s^2 - 10s + 100)$

14. $2(x - 4)(x^2 + 4x + 16)$

_____

_____

Name _____ Date _____ Class _____

The Pythagorean Theorem, $a^2 + b^2 = c^2$, can be used to find the length of the hypotenuse of a right triangle given the length of the two legs. The figure below shows that a right triangle can be created between any two points.

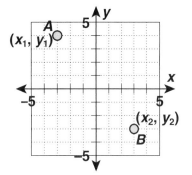 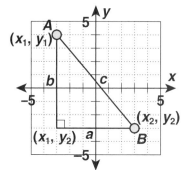

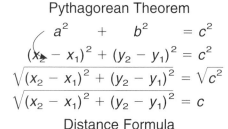

Pythagorean Theorem

$$a^2 \quad + \quad b^2 \quad = c^2$$
$$(x_2 - x_1)^2 + (y_2 - y_1)^2 = c^2$$
$$\sqrt{(x_2 - x_1)^2 + (y_2 - y_1)^2} = \sqrt{c^2}$$
$$\sqrt{(x_2 - x_1)^2 + (y_2 - y_1)^2} = c$$

Distance Formula

The distance between $A(-3, 4)$ and $B(3, -3)$ can be determined by using the distance formula.

$$d = \sqrt{(x_2 - x_1)^2 + (y_2 - y_1)^2}$$
$$d = \sqrt{(3 - (-3))^2 + (-3 - 4)^2}$$
$$d = \sqrt{(6)^2 + (-7)^2}$$
$$d = \sqrt{36 + 49}$$
$$d = \sqrt{85}$$
$$d \approx 9.22$$

So $\overline{AB}$ is 9.22 units long.

Find the distance between the two points. Round your answer to the nearest hundredth.

1. $A(1, 2)$ and $B(4, 6)$ **2.** $C(-5, 0)$ and $D(5, 0)$ **3.** $E(-2, -2)$ and $F(2, 2)$

_____ _____ _____

4. $G(-5, 4)$ and $H(-8, -3)$ **5.** $J(6, -1)$ and $K(3, -6)$ **6.** $L(0, 10)$ and $M(-3, 5)$

_____ _____ _____

7. $N(-2, 1)$ and $P(1, -2)$ **8.** $Q(10, 3)$ and $R(8, -1)$ **9.** $S(-9, 7)$ and $T(-7, 5)$

_____ _____ _____

Saxon Algebra 1

Name _____ Date _____ Class _____

Enrichment
Synthetic Division

93

Synthetic division is a shortcut that can be used when dividing a polynomial by a binomial. In order for synthetic division to work, the divisor must be in the form of $x - c$, that is, a variable minus a constant.

Example: $(x^3 + 6x^2 - x - 30) \div (x - 2)$. Divide.

The value of c is 2.

Write the coefficients of the dividend and the value for c in the upper left corner.

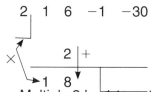

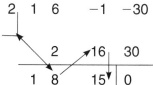

2	1 6 −1 −30

1
Bring down the first coefficient 1 and write it below the horizontal bar.

Multiply 2 by 1 to get 2. Write the product under the next coefficient and add.

Repeat the steps (multiply, write the product under the next coefficient and add) with the remaining numbers.

The quotient is $x^2 + 8x + 15$.

Use synthetic division to find each quotient.

1. $(4x^2 + 19x - 5) \div (x + 5)$

_____

2. $(3y^2 - 5y - 12) \div (y - 3)$

_____

3. $(4a^3 - 3a^2 + 2a - 3) \div (a - 1)$

_____

4. $(5w^3 - 6w^2 + 3w + 14) \div (w + 1)$

_____

5. $(y^3 + 1) \div (y - 1)$ (*Hint*: There are missing terms, fill in the missing terms with 0.)

_____

6. $(2y^5 - 5y^4 - 3y^2 - 6y - 23) \div (y - 3)$

_____

Saxon Algebra 1

Name _____ Date _____ Class _____

Enrichment
Graphing Circles by Completing the Square **104**

Completing the square can be used to graph circles. The general equation for a circle with its center at the origin is $x^2 + y^2 = r^2$, where r is the radius of the circle. The general equation of a circle with its center translated from the origin is $(x - h)^2 + (y - k)^2 = r^2$. An equation representing a circle can be transformed into the sum of two squares.

Example: $x^2 - 14x + y^2 + 6y + 49 = 0$
$(x^2 - 14x + \underline{\quad}) + (y^2 + 6y + \underline{\quad}) = -49$
$(x^2 - 14x + \mathbf{49}) + (y^2 + 6y + \mathbf{9}) = -49 + \mathbf{49} + \mathbf{9}$
$(x - 7)^2 + (y + 3)^2 = 9$
$(x - 7)^2 + (y + 3)^2 = 3^2$

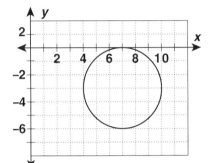

The center of the circle is $(7, -3)$ and the radius is 3.

The circle is shown at the right.

Complete the square on the following equations. Identify the center and radius of the circle and then graph.

1. $x^2 - 8x + y^2 + 2y + 13 = 0$

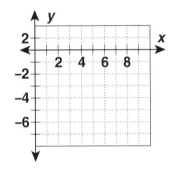

Center: _____

Radius: _____

2. $x^2 + 6x + y^2 + 4y + 12 = 0$

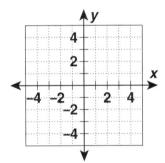

Center: _____

Radius: _____

3. $x^2 + y^2 + 10y - 75 = 0$

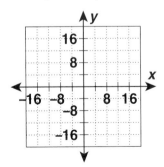

Center: _____

Radius: _____

4. $x^2 - 8x + y^2 - 84 = 0$

Center: _____

Radius: _____

Saxon Algebra 1

Enrichment **105**
Geometric Sequences

A geometric sequence is a sequence in which each term is a product of the previous term and a common ratio, r. For example, 2, 4, 8, 16, ... is a geometric sequence. The common ratio, r is 2. Each term is the product of the previous term and 2.

The common ratio can be determined by finding the quotient of two consecutive terms. In the sequence 1, -4, 16, -64, ... the common ratio is -4 because $\frac{-4}{1} = \frac{16}{-4} = \frac{-64}{16}$.

A geometric sequence has the general form $a_n = a_1 \cdot r^{n-1}$, where n is the term number, and a_1 is the first term in the sequence.

Determine whether each of the following is a geometric sequence.

1. $\frac{1}{3}$, 1, 3, 9, ...

2. 2, 4, 6, 8, ...

3. 1, 1, 2, 3, ...

4. -2, 2, -2, 2, ...

Determine the common ratio for each of the geometric sequences.

5. 5, 15, 45, 135, ...

6. 8, 4, 2, 1, ...

7. -2, 4, -8, 16, ...

8. $\frac{1}{5}$, 1, 5, 25, ...

Write the general form of the geometric sequence.

9. 5, 10, 20, 40, ...

10. -1, -3, -9, -27, ...

11. 100, 10, 1, $\frac{1}{10}$, ...

12. 12, 6, 3, $\frac{3}{2}$, ...

Saxon Algebra 1

Name _____ Date _____ Class _____

 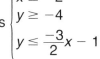
Describe the shaded region of the graph by writing a system of inequalities consisting of three different linear inequalities. To write this system, follow these steps:

1. Determine the slope and *y*-intercept of each line.
2. Write inequalities in slope-intercept form.
3. If the line is solid use either $\leq$ or $\geq$. If the line is dashed, use either $<$ or $>$.
4. If the shaded region is "above" the line use the symbol $>$, and if the shaded region is "below" the line, use the symbol $<$.

For example, the system of inequalities that describes the region below is $\begin{cases} x \geq -2 \\ y \geq -4 \\ y \leq \frac{-3}{2}x - 1 \end{cases}$.

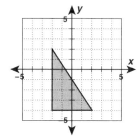

Write a system of inequalities that describes each region.

1.

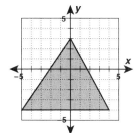

_____

2.

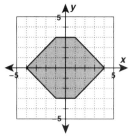

_____

3.

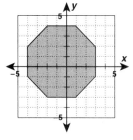

_____

Saxon Algebra 1

Enrichment 115
Graphing Cubic Equations

Just as quadratic equations can be graphed, equations to the third power can also be graphed. To determine the general form of a cubic equation graph the equation $y = x^3$.

Complete the table and plot your points on the graph.

x	$y = x^3$
−2	$(-2)^3 = -8$
−1	$(-1)^3 = -1$
0	$(0)^3 = 0$
1	$(1)^3 = $ _____
2	$(2)^3 = $ _____

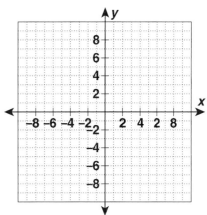

Complete the table and then graph the cubic equations.

1. $y = \frac{1}{3}x^3$

x	y
−3	_____
−1	_____
0	_____
1	
3	_____

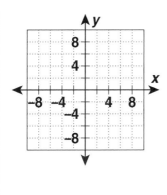

2. $y = -x^3$

x	y
−2	_____
−1	_____
0	_____
1	
2	_____

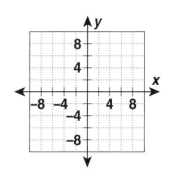

3. $y = x^3 + 2$

x	y
−3	_____
−1	_____
0	_____
1	_____
3	_____

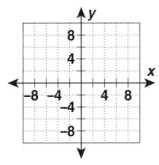

4. $y = x^3 - 2$

x	y
−3	_____
−1	_____
0	_____
1	_____
3	_____

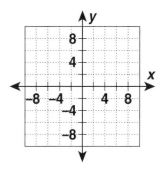

 Saxon Algebra 1

Answer Key

Challenge 1:

The distance Tyrone runs will also include natural numbers, whole numbers, and integers if he runs 8 laps.

Challenge 2

The constants are 5, 2.25, and 3. There are no variables.

$5 + \$2.25 \cdot 3 = \$5 + \$6.75 = \$11.75$

Challenge 3

$$\left(\tfrac{2}{3}\right)^2\left(\tfrac{2}{3}\right)^2\left(\tfrac{9}{4}\right)\left(\tfrac{9}{4}\right) = \left(\tfrac{2}{3}\right)^{2+2}\left(\tfrac{9}{4}\right)^{1+1} =$$

$$\left(\tfrac{2}{3}\right)^4\left(\tfrac{9}{4}\right)^2 = \left(\tfrac{16}{81}\right)\left(\tfrac{81}{16}\right) = 1;$$

$$\left(\tfrac{2}{3}\right)^2\left(\tfrac{2}{3}\right)^2\left(\tfrac{9}{4}\right)\left(\tfrac{9}{4}\right) = \left(\tfrac{4}{9}\right)\left(\tfrac{4}{9}\right)\left(\tfrac{9}{4}\right)\left(\tfrac{9}{4}\right) =$$

$$\left(\tfrac{16}{81}\right)\left(\tfrac{9}{4}\right)\left(\tfrac{9}{4}\right) = \left(\tfrac{144}{324}\right)\left(\tfrac{9}{4}\right) = \left(\tfrac{1296}{1296}\right) = 1$$

Challenge 4

$9 + 12 \div (3 - 1) = 15$

$(3 \cdot 4)^2 - (2^3 + 3)^2 = 23$

Challenge 5

-9

Challenge 6

-109

Challenge 8

about 65 ft/sec

Challenge 9

76

Challenge 10

$0.34, \tfrac{1}{3}, (-0.30 - 0.02), \left(-\tfrac{1}{6} - \tfrac{1}{6}\right)$

Challenge 11

Sample: Simplify the power $(-2)^3 = -8$, and then multiply $-4 \cdot -8 = 32$.

Challenge 12

Sample: Subtraction is the same as adding the opposite. Zero is its own opposite; therefore, it has the same identity as addition.

$$5 - 0 = 5$$

The Community Property does not apply to subtraction because the sign of minuend and subtrahend change.

Challenge 13

$28 \ominus 13 + 18$

Challenge 14

Sample: 1 to 1; $P(A) = 50\%$, $P(\bar{A}) = 50\%$,

and $\dfrac{P(A)}{P(\bar{A})} = \dfrac{0.5}{0.5}$, or 1 to 1

Challenge 15

$mx + xy - 2my - 2y^2$

Challenge 16

5

Challenge 17

$5(n - 2)$

$7 - (3x + 2)$

Challenge 18

$xy(3x - 4)$

Challenge 19

$x = -11$

$x = 11$

$x = 8$

$x = 9$

$x = 21$

Challenge 20

Sample: (3, 3) in quadrant I, $(-1, 3)$ in quadrant II, $(-1, -1)$ in quadrant III, and $(3, -1)$ in quadrant IV

Saxon Algebra 1

Answer Key continued

Challenge 21

$4A = 324$ ft^2

$A = 81$ ft^2 Area of one square

$A = s^2 = 81$ ft^2

$s = \sqrt{81 \text{ ft}^2}$

$s = 9$ ft   Length of one side

$P = 4s$

$P = 4(9 \text{ ft}) = 36$ ft

Challenge 22

Andre's Bank Account

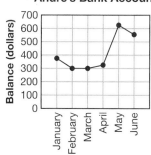

Challenge 23

$3x - 6 = 9$

$3x = 15$

$x = 5$

Challenge 24

4.41; 0.35 is a little more than $\frac{1}{3}$, and

$\frac{1}{3} \cdot 12.6 = 4.2$. Since 4.41 is close to 4.2,

the answer is reasonable.

Challenge 25

0, 6, and -1; Sample: These values will make the respective denominators equal to 0, and division by 0 is undefined.

Challenge 26

$x = -125$

Challenge 28

$8(2x - 1) + 15 + 7x = 4(3x + 5) + 5x - 1$

$\frac{1}{4} \cdot 8(2x - 1) + \frac{1}{4} \cdot 15 + \frac{1}{4} \cdot 7x$

$= \frac{1}{4} \cdot 4(3x + 5) + \frac{1}{4} \cdot 5x - \frac{1}{4} \cdot 1$

$2(2x - 1) + \frac{15}{4} + \frac{7}{4}x = 3x + 5 + \frac{5}{4}x - \frac{1}{4}$

$4x - 2 + \frac{15}{4} + \frac{7}{4}x = 3x + 5 + \frac{5}{4}x - \frac{1}{4}$

$\frac{23}{4}x + \frac{7}{4} = \frac{17}{4}x + \frac{19}{4}$

$\frac{6}{4}x = \frac{12}{4}$

$x = 2$

Challenge 29

$A = \pi r^2$ $\dfrac{\sqrt{81}}{\sqrt{\pi}} = r$

$\dfrac{A}{\pi} = r^2$ $\dfrac{9}{\sqrt{\pi}} = r$

$\sqrt{\dfrac{A}{\pi}} = \sqrt{r^2}$

$\dfrac{\sqrt{A}}{\sqrt{\pi}} = r$

Challenge 30

See Additional Answers.

x	−1	0	1	2	3
y	3	0	−1	0	3

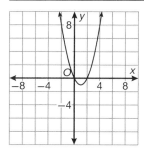

Challenge 31

5

Answer Key continued

Challenge 32

$$\frac{b^3 c^2}{a^6}$$

Challenge 33

$$\frac{2}{9} \cdot \frac{3}{8} \cdot \frac{4}{8} = \frac{1}{24}$$

Challenge 34

816; Sample: The sequence is 6, 12, 18, . . . , 96. Since $96 \div 6 = 16$, there are 16 terms in the sequence. Students should use the formula with $n = 16$, $a_1 = 6$, and $a_n = 96$.

Challenge 35

The graph is a vertical line passing through each point. The x-intercept is 5. There is no y-intercept. The equation is $x + 0y = 5$.

Challenge 36

1:6.25

Challenge 37

$$\frac{\frac{3.5 \times 10^7}{2 \times 10^4} \times \frac{8.1 \times 10^4}{1.5 \times 10^4}}{(1.6 \times 10^7) \times (7.4 \times 10^4)} \quad 7.981 \times 10^{-9}$$

Challenge 38

Sample:
$12g^4 h^8 k^4 - 30g^5 h^7 k^3 + 18g^7 h^9 k^2 - 6g^4 h^{10} k^6$; $(6g^4 h^7 k^2)(2hk^2 - 5gk + 3g^3 h^2 - h^3 k^4)$

Challenge 39

$$\frac{b^2 c^5 d}{a^2} - \frac{5ac}{b^4} + \frac{4b^3 c^7}{a^3 d} - \frac{20c^3}{b^3 d^2}$$

Challenge 40

$$(x^{-3} y^{-2} z^{-1})^{-3}$$

1

Challenge 41

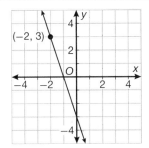

Challenge 42

Paying tutors: $5000; new facility: $10,000; parent-student program: $2400; materials: $2600

Challenge 43

$\frac{1}{2}$; The expression is undefined when $x = 1$ or $x = -1$.

Challenge 44

Car A traveled at a faster speed because it has the greater numerator in the formula for speed.

Challenge 45

$$|2x - 5| \geq (7 + x)^2$$

$$2(x^3 - 1)^2 \leq \frac{x}{2} + 1$$

Challenge 46

30 in.

$$(8 \cdot 3375)^{\frac{1}{3}}$$

$$8^{\frac{1}{3}} \cdot 3375^{\frac{1}{3}}$$

$$8^{\frac{1}{3}} \cdot 3375^{\frac{1}{3}} = 2 \cdot 15 = 30 \text{ in.}$$

Challenge 47

about 1236%

about 0.6%

Saxon Algebra

Answer Key continued

Challenge 48

Sample: I can add up the values higher than my estimate, and subtract my estimate from them. Then I can add up the lower values, and subtract my estimate from them. The sum of the differences above and below my estimate should be equal.

Challenge 49

$y = 4x$; Sample: $(-3, -12)$

Challenge 50

An inequality is appropriate with a restriction of the domain being positive integers.

Challenge 51

Sample: $\dfrac{3x + 1}{x(x - 2)(x + 3)}$

Challenge 52

$x = 22$

Challenge 53

$84x^2$

Challenge 54

box-and-whisker plot, bit, software

Challenge 55

$(2, 2)$ and $(-2, -2)$

Challenge 56

See student work.

Challenge 57

$442x^3y^3z^3$

Challenge 58

$m^2 - m - 12$

$15p^2 - 28p - 32$

$6a^3 - a^2 + 12a + 7$

Challenge 59

The graphs of $y = x + 2$ and $2y - 2x = 4$ are the same line. They have the same slope and the same y-intercept.

Challenge 60

$(x + 8)(x - 8)$

$(2x + 5)(2x - 5)$

$(3a + 4b)(3a - 4b)$

$(x + 7)^2$

$(2a + 3)^2$

$(3x + 7y)^2$

Challenge 61

$2gh\sqrt[3]{h}$

$2z\sqrt[5]{y^3z^3}$

Challenge 62

See student work.

Challenge 63

$x = 2$, $y = -1$, $z = 3$

Challenge 64

$z = \dfrac{4y}{x}$; $z = 16$

Challenge 65

Sample: The slopes of $\overline{QR}$ and $\overline{ST}$ are both -3, so they are parallel. The slopes of $\overline{RS}$ and $\overline{TQ}$ are both $\frac{1}{3}$, so they are parallel. The slope of $\overline{QS}$ is 2, and the slope of $\overline{RT}$ is $-\frac{1}{2}$. These are negative reciprocals of each other, so the diagonals are perpendicular. *QRST* is a rhombus.

Answer Key continued

Challenge 66

false

true

true

false

Challenge 67

Sample: $-4x + y = -1$
$\quad\quad\quad 4x - y = 2$

Sample: $x + 2y = 8$
$\quad\quad\quad \frac{1}{2}x + y = 4$

Sample: $2x + 3y = 3$
$\quad\quad\quad x - y = 4$

Challenge 68

about 106 people

Challenge 69

12

$9x$

$4x^3$

Challenge 70

$0.3x \leq 7.8$; $x \leq 26$ grams

Challenge 71

Sample: An easy test may result in high grades; a difficult test in low grades.

Sample: a discount swimsuit

Sample: A particular player may be popular with the fans.

Challenge 72

$(x - 14)(x + 12)$

Challenge 73

$3 < x \leq 5$

Challenge 74

$2|x - 3| + 4 = 14$; $\{8, -2\}$

Challenge 75

$s(3p - 5s)(3p - 5s)$

$k(2k + 1)(3k + 1)$

$3(2m - 5)(5m - 2)$

$ab(3a - 4b)(3a - 4b)$

$3x(2x + 3)(7x - 3)$

Challenge 76

$9x^2 - 5$

Challenge 77

$b > \dfrac{7}{10}$

$m < 1$

Challenge 78

Sample: It has vertical asymptotes with equations $x = 5$ and $x = -2$. So, the graph will have three pieces. It has x-intercepts at -1 and 3 and a y-intercept at 0.3. As $|x|$ gets larger, y approaches $\dfrac{x^2}{x^2}$ or 1. So, there is a horizontal asymptote at $y = 1$.

Challenge 79

$(x - y)(z + 2)$

$(1 + p)(q - 3)$

Challenge 80

Sample: This particular outcome will occur about $p \cdot n$ times in the next n trials.

Challenge 81

Yes

No

No

Saxon Algebra

Answer Key continued

Challenge 82

$c \leq 60$ OR $c < 40$

Challenge 83

$2x^2y(2xy + 9)(2xy - 9)$

Challenge 84

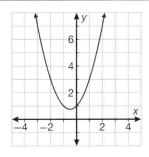

Challenge 85

$(3x)^2 + (4x)^2 = (5x)^2$, where x is a nonzero whole number

Challenge 86

$3\sqrt{2} + \sqrt{290} + 2\sqrt{41}$; 34.08 units

Challenge 87

$(5x^2 + 4y)(6x^3 - 7y^2)$

Challenge 88

$\dfrac{(x + 3)}{(x + 6)}$

Challenge 89

321.5 feet squared

Challenge 90

$\dfrac{3x^9y + x^2 + y}{x^3(x^2 + y)^2}$

Challenge 91

Sample: The solution will be the overlap of the two graphs on a number line,
$-6 < x < -4$

Challenge 92

$\dfrac{x(x - 3)}{x + 2}$

Challenge 93

$(2x + y)$

Challenge 94

Sample: $\left| \dfrac{x}{4} - 2 \right| + 15 = 12$

Challenge 95

$mpz(z - 1)(z - p)$

Challenge 96

2 seconds

Challenge 97

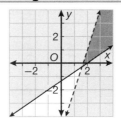

Challenge 98

All these equations are products of squares. They have only one solution.

Challenge 99

$\dfrac{-(m - 3)(m + 5)}{4m^2 - 5m + 4}$

$\dfrac{1}{b + 4}$

Challenge 100

$(1, 0)$

Challenge 101

Sample: $x^2 > 0$

Challenge 102

$x^2 = 225$; 15 inches by 15 inches

Challenge 103

$2\sqrt[3]{4}$

$\dfrac{\sqrt[3]{9}}{3}$

Answer Key continued

Challenge 104

$\{1, -4\}$

Challenge 105

$\frac{3}{4}, \frac{9}{16}$; $\text{Stage}_n = \left(\frac{3}{4}\right)^{n-1}$

Challenge 106

$x = 20$

Challenge 107

Sample: Because there cannot be a negative value for y, the points that fall below the x-axis are "flipped," or reflected over the x-axis.

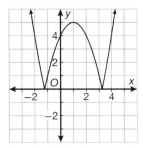

Challenge 108

12.5 grams

Challenge 109

Sample: $\begin{cases} y \geq 1 \\ y \leq 1 \\ x \geq 1 \\ x \leq 1 \end{cases}$

Challenge 110

See student work.

Challenge 111

8 ways: (President, Vice President)(Ann, Charlie), (Ann, Denise)(Bob, Denise)(Charlie, Ann), (Charlie, Denise)(Denise, Ann), (Denise, Bob), (Denise, Charlie)

Challenge 112

$\{(1, 3), (-1, 3)\}$

Challenge 113

$\pm i$

Challenge 114

yes; Sample: Graphs of odd radicals fall on both sides of the y-axis. Graphs of even radicals fall only to the right of the y-axis.

Challenge 115

ends go in opposite directions

ends go in the same direction

ends go in opposite directions

ends go in the same direction

Challenge 116

$6730.10

Challenge 117

$\tan B$, $\cot A$

Challenge 118

$\frac{10}{286} = \frac{5}{143}$

Challenge 119

Sample: The number of pairs of rabbits multiplies every month.

Challenge 120

$\frac{1}{4}$

Answer Key continued

Enrichment 3: The Power of Pi

3.14159265358979323846

1. 81
2. 64
3. 41
4. 625
5. 49
6. 512
7. 256
8. 75
9. 343
10. 625
11. 128
12. 729
13. 387
14. 169
15. 633
16. 52
17. 243
18. 128
19. 864
20. 16

Enrichment 4: Order of Operations

1. 1
2. 125
3. 2
4. 9
5. Undefined
6. 8

Enrichment 7: Digits

2. $3 \cdot \dfrac{2}{1} - 4$
3. $4 - 3 + 2 \cdot 1$

4. $(3 - 2) \cdot \dfrac{4}{1}$
5. $3 \cdot 1 + 4 - 2$
7. $4 + 3 \cdot 1^2$
8. $(4 - 2) \cdot (3 + 1)$
9. $(4 + 1 - 2) \cdot 3$
10. $4 \cdot 3 - \dfrac{2}{1}$
11. $2^3 + (4 - 1)$
12. $2^3 + (4 \cdot 1)$
13. $2^4 - (3 \cdot 1)$
14. $(4 + 3) \cdot \dfrac{2}{1}$
15. $(4 + 2 - 1) \cdot 3$
16. $(3 + 2 - 1) \cdot 4$
17. $(4 + 2) \cdot 3 - 1$
18. $2^4 + (3 - 1)$
19. $2^4 + (3 \cdot 1)$
20. $(3 + 2) \cdot \dfrac{4}{1}$

Enrichment 9: Multiply It Out

1. -6, T
2. -1, I
3. 11, E
4. 3, R
5. 10, U
6. 1, S
7. 9, Y
8. -8, B
9. -7, O
10. 8, D
11. -11, P
12. 0, V

Distributive Property

Answer Key continued

Enrichment 11: All Kinds of Numbers

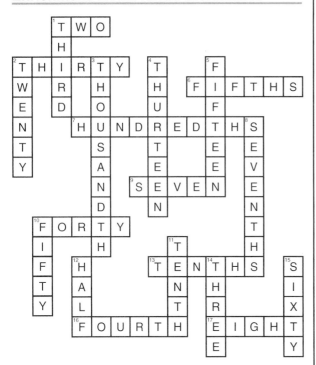

Enrichment 13: Good Luck Squares

50	17	31	80	42	99	69	300	18	115	46	91	63
72	94	65	89	550	10	8	121	97	950	59	150	73
39	615	54	78	16	106	32	112	15	88	19	825	29
125	60	215	211	225	117	377	76	9	82	105	52	85
22	800	81	1	12	250	21	47	196	25	13	115	45
815	4	114	37	116	6	53	7	500	101	525	36	68
51	400	43	113	93	58	40	325	650	715	71	11	28
750	49	725	23	111	144	925	64	74	33	315	169	61
86	98	14	625	900	107	256	109	30	100	5	700	56
57	75	83	110	102	415	20	90	915	95	350	108	77
600	70	66	24	450	48	25	425	850	26	104	84	38
34	87	44	92	79	515	35	103	67	96	55	41	27

Enrichment 17: Number Puzzles

1. $\sqrt{17}$

2. -5

3. $\dfrac{96}{4}$

4. 300

5. 6.5

6. -4

7. $\dfrac{2}{3}$

Enrichment 20: Coordination

Greatest Common Factor

Enrichment 23: Magic Square

8	−5	−6	5
−3	2	3	0
1	−2	−1	4
−4	7	6	−7

The magic sum is 2.

Enrichment 25: Onto Functions

1. Not onto

2. Onto

3. Onto

4. Not onto

Enrichment 26: Perfect Numbers

33,550,336

Enrichment 28: Diophantine Equations

1. a. $y = 3 - \dfrac{3}{4}x$

   b. 4; So the result is an integer.

   c. (4, 0)

2. a. $y = -3 + \dfrac{2}{3}x$

   b. 3; So the result is an integer.

   c. (6, 1)

3. (30, 10)

4. (15, 75)

5. (60, 20)

6. (7, 14)

7. (5, 5)

Saxon Algebra 1

Answer Key continued

8. (4, 5)

9. (3, 7)

10. (2, 3)

Enrichment 30: Profit-Loss-Revenue

1. $s = 45n$

2. $c = 200 + 25n$

3.
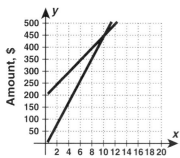

Number of Compact-Disc Players

4. $s = c = 450$

5. $n = 10$

6. $0 \leq n < 10$ or $0 \leq n \leq 9$

7. $n \geq 11$ or $n > 10$

Enrichment 31: Proportions in Paint

1. a. 4:3:12

   b. 10.5 oz maroon, 42 oz deep gold

2. 12 oz thalo green, 26.25 oz magenta

3. 198 g black, 50.4 g blue

Enrichment 42: Fraction, Decimal, and Percent Mania

Boy do I have problems!

Enrichment 44: Intercept–Intercept Form

−4

Sample answer: Use cross products.

$-\dfrac{3}{2}$

1. $a = 1$, $b = 6$, $m = -6$

2. $a = 1$, $b = -1$, $m = 1$

3. $a = 1$, $b = 1$, $m = -1$

4. $a = 2$, $b = 6$, $m = -3$

5. $a = 2$, $b = 2$, $m = -1$

6. $a = 2$, $b = -1$, $m = \dfrac{1}{2}$

Enrichment 45: Triangle Inequalities

1. Obtuse

2. Right

3. Acute

4. Obtuse

5. Right

6. Acute

7. Obtuse

8. Acute

9. $10 \leq x < 12$

10. $x > 11$ or $x < 15$

Enrichment 58: A Maze of Multiplication

The picture shaded is an airplane.

1. $x^2 + 3x$

2. $x^2 + 8x + 7$

3. $x^2 + 2x - 15$

4. $2x^2 - 8x$

5. $8x - 8x^2$

6. $x^2 - 4x - 21$

7. $2x^2 - 11x - 6$

8. $27x - 3x^2$

9. $15x^2 + 2x - 1$

10. $12x^2 + 18x + 6$

11. $x^2 - 64$

12. $45x - 20x^2$

13. $14x - 6$

14. $x^2 - 13x + 30$

15. $30 + 7x - 2x^2$

16. $9x^2 - 49$

Saxon Algebra 1

Answer Key

17. $10x - 4x^2$

18. $40x^2 - 43x - 6$

19. $9x^2 - 4$

20. $7x^2 - 32x - 15$

Enrichment 60: Pascal's Triangle

4 and 6

1, 7, 21, 35, 35, 21, 7, 1

1, 3, 3, 1

1. $x^4 + 4x^3y + 6x^2y^2 + 4xy^3 + y^4$

2. $x^5 + 5x^4 + 10x^3 + 10x^2 + 5x + 1$

3. $x^4 + 12x^3 + 54x^2 + 108x + 81$

4. $x^3 + 6x^2 + 12x + 8$

5. $x^9 + 9x^8 + 36x^7 + 84x^6 + 126x^5 + 126$
$x^4 + 84x^3 + 36x^2 + 9x + 1$

6. $x^5 + 10x^4y + 40x^3y^3 + 80x^2y^3 + 80xy^4$
$+ 32y^5$

Enrichment 64: Other Types of Variation

1. $V = k\dfrac{T}{P}$

2. a. $0.0022\dfrac{Latm}{K}$

b. $V = 1.408$ L

c. 0.352 atm

3. a. $k = 0.00125$, $V = 5.714$ L

b. A drop of 40 Kelvin, from 320 to 280

Enrichment 67: Using a Matrix to Represent a System of Equations

1. d

2. a

3. b

4. c

5. $\begin{bmatrix} 1 & -5 \\ 2 & -3 \end{bmatrix}\begin{bmatrix} x \\ y \end{bmatrix} = \begin{bmatrix} 0 \\ 7 \end{bmatrix}$

6. $\begin{bmatrix} 4 & 3 \\ 3 & -4 \end{bmatrix}\begin{bmatrix} x \\ y \end{bmatrix} = \begin{bmatrix} 19 \\ 8 \end{bmatrix}$

7. $\begin{bmatrix} 5 & 3 \\ 4 & -5 \end{bmatrix}\begin{bmatrix} x \\ y \end{bmatrix} = \begin{bmatrix} 12 \\ 17 \end{bmatrix}$

8. $\begin{bmatrix} 1 & 1 \\ 1 & -1 \end{bmatrix}\begin{bmatrix} x \\ y \end{bmatrix} = \begin{bmatrix} 7 \\ 9 \end{bmatrix}$

9. $\begin{bmatrix} 12 & -9 \\ 12 & 7 \end{bmatrix}\begin{bmatrix} x \\ y \end{bmatrix} = \begin{bmatrix} 114 \\ 82 \end{bmatrix}$

10. $\begin{bmatrix} 2 & -3 \\ 1 & 3 \end{bmatrix}\begin{bmatrix} x \\ y \end{bmatrix} = \begin{bmatrix} -4 \\ 7 \end{bmatrix}$

11. $\begin{bmatrix} \frac{1}{2} & 1 \\ 1 & \frac{1}{5} \end{bmatrix}\begin{bmatrix} x \\ y \end{bmatrix} = \begin{bmatrix} 12 \\ 10 \end{bmatrix}$

12. $\begin{bmatrix} \frac{2}{3} & \frac{1}{3} \\ \frac{1}{4} & \frac{3}{4} \end{bmatrix}\begin{bmatrix} x \\ y \end{bmatrix} = \begin{bmatrix} -9 \\ 16 \end{bmatrix}$

13. $\begin{bmatrix} 1.2 & -1.6 \\ -0.8 & 0.2 \end{bmatrix}\begin{bmatrix} x \\ y \end{bmatrix} = \begin{bmatrix} 2.4 \\ -1.2 \end{bmatrix}$

Enrichment 72: Factoring Fun

Answer: 2748

1. $(x - 2)(x - 4)$

2. $(x - 2)(x + 3)$

3. $(x - 1)(x - 4)$

4. $(x + 2)(x - 1)$

5. $(x + 2)(x - 4)$

6. $(x - 3)(x + 4)$

7. $(x + 3)(x - 4)$

8. $(x + 1)(x - 4)$

9. $(x - 2)(x - 1)$

10. $(x - 4)(x - 4)$

11. $(x - 5)(x + 3)$

12. $(x + 2)(x + 3)$

13. $(x - 2)(x - 3)$

14. $(x + 5)(x - 4)$

15. $(x - 2)(x + 1)$

16. $(x + 5)(x - 1)$

17. $(x - 2)(x + 4)$

18. $(x - 2)(x + 5)$

Enrichment 75: Fourth Degree Trinomials

1. $(4d^2 - d + 1)(4d^2 + d + 1)$

2. $(p^2 + p + 1)(p^2 - p + 1)$

 Saxon Algebra

Answer Key continued

3. $(2x^2 + 3x - 1)(2x^2 - 3x - 1)$

4. $(2x^2 - 5xy + 4y^2)(2x^2 + 5xy + 4y^2)$

5. $(3r^2 + 2rs + 5s^2)(3r^2 - 2rs + 5s^2)$

6. $(2a^2 - 5ac + 5c^2)(2a^2 + 5ac + 5c^2)$

Enrichment 82: The Greatest Possible Error

1. 0.05 m

2. 0.05 g

3. 0.005 L

4. 0.005 miles

5. $3.15 \leq l \leq 3.25$

6. $1.45 \leq b < 1.55$

7. $2.745 \leq p < 2.755$

8. $1.745 \leq d < 1.755$

9. minimum area $= 11.5(7.5) = 86.25$ ft^2; maximum area $= 12.5(8.5) = 106.25$ ft^2

86.25 ft$^2 \leq$ area < 106.25 ft^2

10. minimum area $= 10.335^2 = 106.812$ cm^2; maximum area $= 10.345^2 = 107.019$ cm^2

106.812 cm$^2 \leq$ area < 107.019 cm^2

Enrichment 83: Sum and Differences of Cubes

1. $(r - s)(r^2 + rs + s^2)$

2. $(x + y)(x^2 - xy + y^2)$

3. $(x + 2)(x^2 - 2x + 4)$

4. $(n - 4)(n^2 + 4n + 16)$

5. $(2y + 3)(4y^2 - 6y + 9)$

6. $p(q - 4)(q^2 + 4q + 16)$

7. $m^3 - 1$

8. $8 + 27t^3$

9. $b^3 - 4^3$

10. $x^3 + 7^3$

11. $8y^3 - 1$

12. $27 + 8t^3$

13. $s^3 + 1000$

14. $2x^3 - 128$

Enrichment 85: Derive the Distance Formula from the Pythagorean Theorem

1. 5

2. 10

3. 5.66

4. 7.62

5. 5.83

6. 5.83

7. 4.24

8. 4.47

9. 2.83

Enrichment 93: Synthetic Division

1. $4x - 1$

2. $3y + 4$

3. $4a^2 + a + 3$

4. $5w^2 - 11w + 14$

5. $y^2 + y + 1 + \dfrac{2}{(y - 1)}$

6. $2y^4 + y^3 + 3y^2 + 6y + 12 + \dfrac{13}{y - 3}$

Enrichment 104: Graphing Circles by Completing the Square

1.

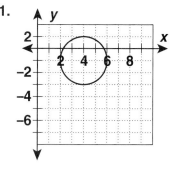

$(4, -1)$
2

Saxon Algebra 1

Answer Key continued

2.

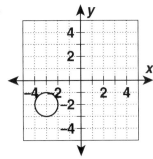

$(-3, -2)$
1

3.

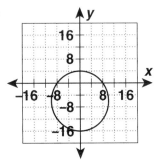

$(0, -5)$
10

4.

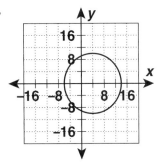

$(4, 0)$
10

Enrichment 105: Geometric Sequences

1. Yes

2. No

3. No

4. Yes

5. 3

6. $\frac{1}{2}$

7. -2

8. 5

9. $a_n = 5 \cdot 2^{n-1}$

10. $a_n = -1 \cdot 3^{n-1}$

11. $a_n = 100 \cdot \left(\frac{1}{10}\right)^{n-1}$

12. $a_n = 12 \cdot \left(\frac{1}{2}\right)^{n-1}$

Enrichment 109: Describing Geometric Regions with a System of Inequalities

1. $\begin{cases} y \le \frac{3}{2}x + 3 \\ y \le -\frac{7}{4}x + 3 \\ y \ge -4 \end{cases}$

2. $\begin{cases} y \le 3 \qquad\quad y \ge -3 \\ y \ge x - 4 \qquad y \ge -x - 4 \\ y \le -x + 4 \quad y \le x + 4 \end{cases}$

3. $\begin{cases} x \ge -4 \quad x \le 3 \quad y \ge -x - 5 \\ y \le 4 \qquad y \ge -3 \quad y \le x - 4 \\ y \ge -x + 5 \qquad\qquad y \le x + 6 \end{cases}$

Enrichment 115: Graphing Cubic Equations

1

8

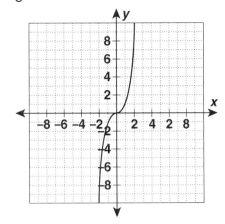

Saxon Algebra

1. $-9, -\frac{1}{3}, 0, \frac{1}{3}, 9$

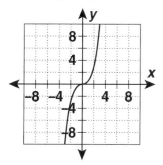

2. 8, 1, 0, −1, −8

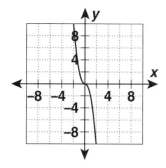

3. −25, 1, 2, 3, 29

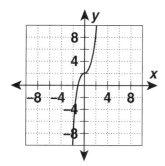

4. −29, −3, −2, −1, 25

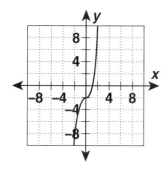